Alternative Fuel Vehicles

Alternative Fuel Vehicles gives full coverage of all associated qualifications and awards in the emerging field of alternative fuels. It is an essential introduction to the ever-growing demand for vehicles that operate using non-conventional fuels.

This first book on AFVs endorsed by the IMI begins with an overview of the subject, ideal for beginners, before outlining what is meant by alternative fuels, why they are necessary, and why climate change and associated legislation are key drivers. Details of how alternative fuels are made, the supply infrastructure, and how these vehicles work are all included. A chapter on fuel cells introduces learners to the use of hydrogen, and one on engines and engine management includes coverage of combustion as an aid to understanding why changing the type of engine fuel is complex. Some basic engine technology is included to help readers new to the subject. Real-life case studies and examples are used to illustrate different technologies in current use, and to speculate on new developments. This book is an ideal companion to any unit of study on alternative fuel, but will also be of interest to working technicians and keen amateurs.

Tom Denton is the leading UK automotive author with a teaching career spanning from lecturer to head of automotive engineering in a large college. His wide range of automotive textbooks published since 1995 are bestsellers and led to his authoring of the Automotive Technician Training multimedia system that is in common use in the UK, USA and several other countries.

Alternative Fuel Vehicles

Tom Denton

Routledge
Taylor & Francis Group
LONDON AND NEW YORK

INSTITUTE OF THE MOTOR INDUSTRY

First published 2018
by Routledge
2 Park Square, Milton Park, Abingdon, Oxon OX14 4RN

and by Routledge
711 Third Avenue, New York, NY 10017

Routledge is an imprint of the Taylor & Francis Group, an informa business

British Library Cataloguing-in-Publication Data
A catalogue record for this book is available from the British Library

Library of Congress Cataloging-in-Publication Data
Names: Denton, Tom, author.Title: Alternative fuel vehicles / Tom Denton.
Description: Boca Raton : Taylor & Francis, 2018. | Includes bibliographical references and index.
Identifiers: LCCN 2017057731 (print) | LCCN 2018000583 (ebook) | ISBN 9781315512242 (Adobe PDF) |
    ISBN 9781315512235 (ePub) | ISBN 9781315512228 (Mobipocket) | ISBN 9781138201316 (pbk. : alk. paper) |
    ISBN 9781138503700 (hardback : alk. paper) | ISBN 9781315512259 (ebook : alk. paper)
Subjects: LCSH: Alternative fuel vehicles.
Classification: LCC TL216.5 (ebook) | LCC TL216.5 .D46 2018 (print) | DDC 629.22/9–dc23
LC record available at https://lccn.loc.gov/2017057731

ISBN: 978-1-138-50370-0 (hbk)
ISBN: 978-1-138-20131-6 (pbk)
ISBN: 978-1-315-51225-9 (ebk)

Typeset in Univers by
Servis Filmsetting Ltd, Stockport, Cheshire

Contents

Preface

In this book you will find lots of useful and interesting information about Alternative Fuel Vehicles (AFVs). It is the fifth in the 'Automotive Technology: Vehicle Maintenance and Repair' series:

▶ Automobile Mechanical and Electrical Systems
▶ Automobile Electrical and Electronic Systems
▶ Automobile Advanced Fault Diagnosis
▶ Electric and Hybrid Vehicles
▶ Alternative Fuel Vehicles
▶ Automated Driving Systems.

Ideally, you will have studied the mechanical book, or have some experience, before reading this one. If not, it does include some revision sections so don't worry!

This book concentrates on alternative fuels but does not include detailed electric vehicle technology as this is all in a separate book. Some aspects of climate change and legislation are covered to set the context for the developments of alternative fuels. Case studies are used to show real systems in use. I have not covered aftermarket conversion kits as these are becoming less common.

Comments, suggestions and feedback are always welcome at my website:

www.automotive-technology.org

On this site, you will find lots of **free** online resources to help with your studies. Check out Chapter 7 for more information about the amazing resources to go with this and my other books. These resources work with the book, and are ideal for self-study or for teachers helping others to learn.

Good luck with your studies, and I hope you find automotive technology as interesting as I still do.

Acknowledgements

Over the years many people have helped in the production of my books. I am therefore very grateful to the following companies who provided information and/or permission to reproduce photographs and/or diagrams:

AA
AC Delco
ACEA
Alpine Audio Systems
Audi
Autologic Data Systems
BMW UK
Brembo brakes
C&K Components
Citroën UK
Clarion Car Audio
Continental
CuiCAR
Dana
Delphi Media
Eberspaecher
Fluke Instruments UK
Flybrid systems
Ford Motor Company
FreeScale Electronics
General Motors
GenRad
haloIPT (Qualcomm)
Hella
HEVT
Honda
Hyundai

Institute of the Motor
    Industry
Jaguar Cars
Kavlico
Loctite
Lucas UK
LucasVarity
Mahle
Mazda
McLaren Electronic
    Systems
Mennekes
Mercedes
Mitsubishi
Most Corporation
NASA
NGK Plugs
Nissan
Oak Ridge National Labs
Peugeot
Philips
PicoTech/PicoScope
Pierburg
Pioneer Radio
Porsche
Renesas
Robert Bosch Gmbh/Media

Rolec
Rover Cars
Saab Media
Scandmec
SMSC
Snap-on Tools
Society of Motor Manufacturers
    and Traders (SMMT)
Sofanou
Sun Electric
T&M Auto-Electrical
Tesla Motors
Thrust SSC Land Speed Team
Toyota
Tracker
Tula
Unipart Group
Valeo
Vauxhall
VDO Instruments
Volkswagen
Volvo Cars
Volvo Trucks
Wikimedia
ZF Servomatic

If I have used any information, or mentioned a company name that is not listed here, please accept my apologies and let me know so that it can be rectified as soon as possible.

CHAPTER 1

Introduction

1.1 What is an alternative fuel?

The general definition of an 'alternative fuel' for a vehicle is one that is not produced from crude oil. However, this explanation can be blurred as some alternative fuels are mixed with petroleum base fuels and, further, some are derived from oil, but used in a different way. More about this later.

> **Definition**
> **Alternative fuel**: Not produced from fossil deposits such as crude oil, coal or shale.

Fuels are generally thought of as liquids because the two main fuels used for automotive applications are liquids at room temperature:

1 Petrol/Gasoline
2 Diesel.

In 2016, the worldwide average demand was over 90 million barrels[1] of oil and liquid fuels per day. That is about 35 billion barrels a year! Production reached 97 million barrels per day in late 2015, and a demand for almost 100 million barrels per day is expected in the next few years (BP, 2017).

> **Key fact**
> In 2016 the worldwide average demand for oil and liquid fuels was over 90 million barrels per day.

There are other alternative fuels but the ones that will be examined in more detail in this book are:

1 Ethanol (Bio-alcohol)
2 Methanol
3 Biodiesel
4 Liquified natural gas (LNG)
5 Compressed natural gas (CNG)
6 Liquified petroleum gas (LPG)
7 Hydrogen
8 Solar
9 Dimethyl ether fuel.

Some of these fuels are used to make electricity. The method of conversion will be examined but not the additional electric vehicle technology. Please refer to *Hybrid and Electric Vehicles* (Denton, 2016) for more details on these areas.

Figure 1.1 Crude oil being pumped from the ground[2]

Safety first

All fuels are highly flammable, some are corrosive and some are stored under very high pressure – take care!

Key fact

European Commission directive requires Member States to adopt national policy.

1.2 Infrastructure and regulations

In January 2013, the European Commission proposed a directive requiring Member States to adopt national policy frameworks for developing the market for alternative fuels and to ensure that minimum infrastructure is set up for their supply in road and water-borne transport:

▶ Each Member State should ensure the establishment of a defined minimum number of recharging points for electric vehicles by the end of 2020 (at least 10% of them publicly accessible). Ports should be equipped with shore-side electricity supply for vessels by end-2015.

▶ Hydrogen refuelling points should be set up in sufficient number (no further than 300 km apart) to allow hydrogen vehicles to circulate throughout the territory (by 2020 in Member States where this technology has already been introduced).

▶ LNG supply should be available for navigation along the core Trans-European Transport (TEN-T) network in maritime ports (2020) and inland ports (2025), and LNG refuelling points should sustain heavy-vehicle road transport along the core network (refuelling points at least every 400 km by 2020).

▶ By end-2020, Member States should ensure sufficient CNG refuelling points are set up (at least every 150 km) to support CNG vehicles across the EU. This proposal would also require harmonisation of technical specifications of alternative fuels, and common standards for refuelling and electric charging systems, and more information to consumers on compatibility of fuels and vehicles.

Notwithstanding the results of Brexit negotiations (ongoing at the time of writing . . .), the UK has put forward regulations to implement the requirements of Directive (2014/94/EU)[3] of the European Parliament and of the Council of 22 October 2014. This is about establishing a common framework of measures for the deployment of alternative fuels infrastructure. The purpose of the Directive is to minimise dependence on oil and to mitigate the environmental impact of transport. The significant points for the context we are working within are listed below. For alternative fuel infrastructure deployed or renewed after 17 November 2017, the requirements are specified in the Schedule to the Regulations.

▶ Normal or high-power recharging points for electric vehicles comply with the minimum technical standards for socket outlets or vehicle connectors.
▶ Refuelling points supplying hydrogen meet with a technical standard in relation to their connectors for motor vehicles.
▶ Data indicating the geographic location of public recharging or refuelling points, when available, must be accessible to the public on an open and non-discriminatory basis.

The EC Fuel Quality Directive (FQD) (2009/30/EC) defines standards for transport fuels and requires that fuel suppliers meet a 6% reduction in greenhouse gas emissions by 2020, relative to 2010 baseline levels, across all fuel categories. The FQD specifies that ethanol may be blended into petrol up to a limit of 10% by volume.

In the UK the Renewable Transport Fuel Obligation (RTFO) order applies to large suppliers and requires a percentage of the fuel supplied to come from renewable and sustainable sources.

Fuel specs are defined in European standards developed by governments, the oil industry and the car industry working together to make sure that petrol and diesel are suitable for use in the range of different vehicle and engines.

The standard specifications[4] of petrol and diesel in the UK are British Standard (BS) versions of European Standards (EN):

▶ BS EN 228 for petrol
▶ BS EN 590 for diesel.

These regulations first allowed for up to 5% of ethanol to be blended in petrol and 5% biodiesel in diesel so that fuel suppliers could meet the RTFO. At the 5% level, there was no issue of compatibility with car fuel systems and no need to mark pumps to tell customers that the fuel may contain biofuel. In March 2013 the maximum ethanol allowed in petrol increased from 5% to 10%. There may be compatibility issues with some older fuel system components at this level. The updated standard therefore stated that any petrol containing more than 5% ethanol must be clearly labelled on the pump as: *Unleaded petrol 95 E10*. Filling stations supplying E10

Figure 1.2 Fuel pump with a range of options

3

Figure 1.3 Biofuel dispenser for several ethanol and biodiesel blends[5]

also offer an E5 version for non-compatible vehicles.

Over 90% of petrol vehicles on the road are compatible with E10 but 10% therefore are not! The UK government discouraged an early switch to E10 so that the number of incompatible vehicles will reduce as they reach their end-of-life.

> **Key fact**
> Over 90% of petrol vehicles on the road are compatible with E10.

Petrol cars made since 2000 are more likely to be compatible with ethanol, but when E10 is fully introduced vehicle manufacturers and fuel suppliers will have to provide information about vehicle compatibility. If an older petrol/gasoline car is miss-fuelled with a high ethanol version, it is not necessary to drain the tank if the next fill is correct.

Biodiesel is becoming more common, particularly FAME (fatty acid methyl ester, made from oils and greases). This is compatible with modern emissions control systems, and contains oxygen so enhances combustion and reduces emissions of CO, HC and particulate matter. However, like ethanol, this biodiesel has higher solvency so can cause issues with fuel system rubbers at higher blending rates. It can also have increased waxing problems at lower temperatures.

1.3 History

A French Army Captain, called Cugnot, was one of the first to create a machine that could convert the reciprocating motion of a steam piston into rotary motion. A small version of his three-wheeled *fardier à vapeur* (steam dray) was made and used in 1769. A *fardier* was a heavy two-wheeled horse-drawn cart.

In 1770, a full-size version was built to carry a load of 4 tonnes about 5 miles, in 1 hour. The vehicle weighed about 2.5 tonnes, had two wheels at the rear and one at the front where the horses would normally have been! The front wheel supported a steam boiler and the driving mechanism. Sadly, while the steam wagon did work it never quite lived up to the hopes of the designer!

Sometime around 1807, Frenchman Nicéphore Niépce, and his brother Claude, invented an internal-combustion engine, which they called the Pyréolophore. This name was derived from a combination of the Greek words for fire, wind and 'I produce'. The engine used a piston in a cylinder like modern engines. It first used lycopodium powder as a fuel, which was later combined with coal dust. The brothers claimed to have used it to power a boat. Niépce didn't do much more work on the engine but is now credited with inventing photography!

Figure 1.4 A filling station with four alternative fuels for sale: biodiesel (B3), gasohol (E25), neat ethanol (E100) and compressed natural gas (CNG)[6]

1 Introduction

Coal dust was also used by Rudolf Diesel in a series of experimental engine tests in 1892. Fortunately, he changed direction and developed the compression ignition engine, which is now named after him. In the 1970s General Motors researchers experimented with coal dust when looking for alternatives to oil supplied from the Middle East. Coal dust is energy dense, readily available and cheap. However, it is difficult to handle and can tend to explode. It is not now used but coal-to-liquid processes are used for producing synthetic gasoline.

The first commercial vehicle that was able to use ethanol as a fuel was the Ford Model T, produced from 1908 to 1927. It was fitted with a carburettor with adjustable jetting (or in some cases a different carburettor completely), allowing use of gasoline or ethanol, or a combination of both. The Model T was never fitted with a dual fuel change over switch on the dashboard, as some Internet rumours would have us believe!

From around 1935 Ford built trucks and buses equipped with an engine that would run on wood (charcoal) gas. These were known as generator vehicles. If loaded with wood the range of some buses was about 230 miles. The purpose of these vehicle was not to improve the environment, it was to try to become independent of the oil companies.

A Belgian inventor first proposed using transesterification in 1937, to convert vegetable oils into fatty acid alkyl esters and use them as a diesel fuel replacement.

Smaller plants started to produce biodiesel from the 1980s and it is now a significant part of the world economy. One of the first large-scale commercial plants in the United States was established in 1996 in Hawaii. The plant recycled used cooking oil into biodiesel.

Volkswagen had a Bora model running on a methanol fuel cell in the year 2000. Their first liquid hydrogen version was 2001.

Hyundai Motor Company began sales of the Elantra LPI Hybrid in the South Korean domestic market in 2009. The Elantra LPI (Liquefied Petroleum Injected) was the world's first hybrid electric vehicle to be powered by an internal combustion engine built to run on liquefied petroleum gas (LPG) as a fuel.

Honda introduced the Clarity in 2008 and a newer version in 2017. One of the latest innovations is the Toyota Mirai. Both manufacturers' cars use hydrogen fuel cells.

Volvo Trucks are currently (2017/2018) producing LNG engines for their large long-range vehicles. Significant reductions in emissions and improved mileage have been the result.

Some of these technologies are covered in more detail later.

1.4 Developments

Recent changes in both attitude and subsequent legislation seem to suggest that the days of the internal combustion engine are numbered. Diesel cars in cities are currently causing much concern. Even though these

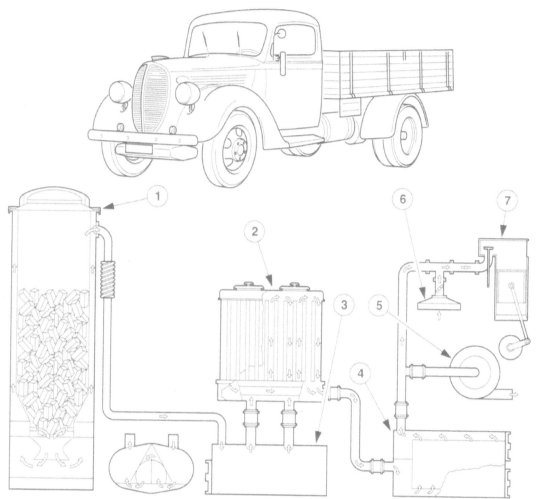

Figure 1.5 Wood gas operation: 1. Gas producer, 2. Gas cooler, 3. Settling container, 4. Filter, 5. Blower, 6. Air filter, 7. Engine. (Source: Ford Motor Company)

cars meet very stringent regulations the sheer numbers moving slowly past schools, for example, results in very poor quality air in localised areas. It will be a few decades yet before we see the demise of the IC engine – and perhaps longer as better alternative fuels are developed and used. New engine combustion technologies are also being developed and examples are discussed later in the book.

One key to the development of viable alternative fuels is energy density (see Table 1.1). Electric vehicle batteries, while improving all the time, have nowhere near the density of liquid fuels. However, electric cars are now available with operating ranges of 200 miles (320 km) or more. This makes these cars practicable for normal use.

Key fact
One key development for viable alternative fuels is high energy density.

Hydrogen, while expensive to produce, has a massive energy density and the use of fuel cell technology means it can be used efficiently. At the time of writing, hydrogen fuel cell cars are limited and very expensive, but this will change.

There are some experimental systems that may come to something. An interesting method uses liquid nitrogen (LN_2) as a method of storing energy. Energy is used to liquefy air, and then LN_2 is produced by evaporation. The liquid nitrogen is exposed to ambient heat in the vehicle and the resulting gas can be used to power a piston or turbine engine. The maximum amount of energy that can be extracted is 213 Watt-hours per kg (Wh/kg) or 173 Wh per litre, in which a maximum of 70 Wh/kg can be utilised in a normal engine. A vehicle with a 350-litre tank could have a range like a gasoline powered vehicle with a 50-litre tank. But note that is seven times larger and at high pressure so there are some technical issues. None the less, the emissions are completely inert.

> **Key fact**
> Liquid nitrogen can be used as a method of storing energy.

A hybrid bus that runs on both diesel and compressed nitrogen has been on trial. It uses a Dearman engine alongside a conventional diesel. Below 20 mph the system warms the nitrogen until it boils and therefore creates enough pressure to drive the engine. Above 20 mph the normal diesel engine takes over.

> **Definition**
> **Dearman engine:** A piston engine driven by the expansion of liquid nitrogen or air, invented by Peter Dearman.[7]

During 2015 to 2016 in the UK, low carbon fuel made up 3% of the total fuel used. This generated significant emission savings equating to taking 1,000,000 cars off the road, according to the Department for Transport (DfT).[8] In the UK this is estimated to contribute £30 million per year to the economy.

The largest feedstock for these fuels was used cooking oil, which was used to produce FAME biodiesel. A small amount was used to create hydrogenated vegetable oil (HVO). Because HVO uses hydrogen as the catalyst (FAME uses methanol), the oxygen is removed and therefore the fuel is more stable and can be stored longer. Scania reports that all its Euro 6 engines in production can be run on HVO.[9] Next generation trucks will run on a wide range of FAME, HVO and diesel.

Notes

1   A barrel of oil is equal to 159 litres, 42 US gallons or 35 imperial gallons.
2   By Tim Evanson – www.flickr.com/photos/23165290@N00/9287130523/,CC BY-SA 2.0
3   http://eur-lex.europa.eu/legal-content/EN/TXT/?uri=CELEX:32014L0094
4   Standards in the USA are arranged by the Environmental Protection Agency (EPA): https://www.epa.gov/gasoline-standards
5   Wikimedia Commons contributors, WAS 2010 8953.jpg, https://commons.wikimedia.org
6   By Mariordo Mario Roberto Duran Ortiz – Own work, CC BY-SA 3.0, https://commons.wikimedia.org
7   http://dearman.co.uk/dearman-technologies/dearman-engine/
8   https://www.gov.uk/government/collections/biofuels-statistics
9   https://www.scania.com/group/en/from-the-leader-in-alternative-fuelsgreen-light-for-hvo-use-in-scania-euro-6-range-3/

CHAPTER 2

The environment

2.1 Introduction

Our planet's natural environment encompasses all naturally occurring living and non-living things. This environment includes the interaction of all living species, climate, weather, and natural resources that affect human survival and economic activity.

The idea of the natural environment can be considered in two ways:

1 Complete ecological units that function as natural systems without massive civilised human intervention, including all vegetation, micro-organisms, soil, rocks, atmosphere, and natural phenomena that occur within their boundaries and their nature.
2 Universal natural resources and physical phenomena that lack clear-cut boundaries, such as air, water, and climate, as well as energy, radiation, electric charge, and magnetism, *not* originating from civilised human activity.

Then there is the built environment, where landscapes have been transformed by human activity. For example, urban developments and agricultural land conversion. Even less extreme things, such as building a mud hut or a photovoltaic system in the desert, change the natural environment into an artificial one.

The changes to the natural environment that humans can cause vary from hardly any impact to a 100% change. Transport in general, does of course have a significant impact.

2.2 Facts and figures

According to the European Automobile Manufacturers Association (ACEA: www. acea.be), there are some 256 million cars on Europe's roads with an average age of about 10 years. Half of new cars sold (2017) are diesel powered, although this is tending to reduce as their impact on local environments becomes more understood.

> **Key fact**
>
> There are over a quarter of a billion cars on Europe's roads (2018).

Alternative fuel vehicles (EVs, CNG, LPG etc.) make up about 6% of the total EU fleet and 4.2% of new car sales. An average new car emits 118.1g CO_2/km.

Key fact

Alternative fuel vehicles make up about 6% of the total EU fleet (2018).

In the third quarter of 2017, registrations of alternative fuel vehicles (AFVs) in the European Union continued to show strong growth, with demand increasing by over 50% compared with the previous year. There were 211,635 alternatively powered cars registered during this period, accounting for 6.2% of total passenger car sales. Electrically chargeable vehicles (ECVs) made up for 1.6% of all cars sold across the EU during the third quarter of the year.

Table 2.1 EU figures relating to vehicle production and use. (Source: ACEA, www.acea.be)

EMPLOYMENT		
Manufacture of motor vehicles (EU28)	2.5 million people = 8.2% of EU employment in manufacturing	2015
Total (EU28 manufacturing, services and construction)	12.6 million people = 5.7% of total EU employment	2015
PRODUCTION		
Motor vehicles (world)	96.1 million units	2016
Motor vehicles (EU28)	19.2 million units = 20% of global motor vehicle production	2016
Passenger cars (world)	77.7 million units	2016
Passenger cars (EU28)	16.5 million units = 21% of global passenger car production	2016
REGISTRATIONS		
Motor vehicles (world)	95.1 million units	2016
Motor vehicles (EU27)	17.0 million units = 18% of global motor vehicle registrations/sales	2016
Passenger cars (world)	77.3 million units	2016
Passenger cars (EU27)	14.6 million units = 19% of global passenger car registrations/sales	2016
Diesel (EU15)	49.9%	2016
Electric (EU15)	1.1%	2016
VEHICLES IN USE		
Motor vehicles (EU28)	294.2 million units	2015
Passenger cars (EU28)	256.1 million units	2015
Motorisation rate (EU28)	573 units per 1,000 inhabitants	2015
Average age (EU25)	10.7 years	2015
TRADE		
Exports (extra-EU28)	€135.4 billion	2016
Imports (extra-EU28)	€45.7 billion	2016
Trade balance	€89.7 billion	2016
ENVIRONMENT		
Average CO_2 emissions (EU28)	118.1g CO_2/km	2016
INNOVATION		
Automobiles & parts sector	€50.1 billion	2015
TAXATION		
Fiscal income from motor vehicles (EU14)	€395.7 billion	2016

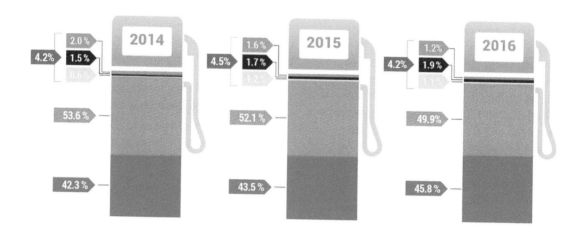

1. Includes battery electric vehicles (BEV), extended-range electric vehicles (EREV), plug-in hybrid electric vehicles (PHEV) and fuel cell electric vehicles (FCEV)
2. Includes full and mild hybrids
3. Includes natural gas vehicles (NGV), LPG-fueled vehicles and ethanol (E85) vehicles

Figure 2.1 New passenger cars in the EU by fuel type. (Source: ACEA, www.acea.be)

Registrations (third quarter, 2017) of battery, hybrid and plug-in electric cars accounted for the highest percentage gains, going up by 60.9%, 59.7% and 53.2% respectively. Demand for new LPG-fuelled cars increased by 38.1% compared with the same period the previous year, and registrations of cars powered by natural gas were 12.9% higher.

These percentage increases are impressive, but they are coming from a low base. When looking at the relative numbers, the market share of alternative fuel vehicles remains small. Only one in 60 cars sold during the third quarter of 2017 was an electrically chargeable one. All alternative technologies combined, including hybrid, fuel cell, ethanol (E85), LPG and natural gas vehicles, accounted for one in 16 new passenger cars registered in the EU.[1]

2.3 Global warming

Global warming, also referred to as climate change, is the observed long term rise in the average temperature of the Earth's climate system, and its related effects. Many different strands of scientific research and evidence show that the climate system is warming. This is now accepted by all but a few most ardent sceptics. Many of the observed changes since the 1950s are unprecedented in the instrumental temperature record, which extends back to the mid-nineteenth century, and in paleoclimate records covering thousands of years. In 2013, the Intergovernmental Panel on Climate Change (IPCC) Fifth Assessment Report concluded that:

It is extremely likely that human influence has been the dominant cause of the observed warming since the mid-20th century.

The main human influence has been the emission of greenhouse gases such as carbon dioxide and methane. Climate model projections in the IPCC report indicated that during the twenty-first century the global surface temperature is likely to rise a further 0.3 to 1.7°C in the lowest emissions scenario, or 2.6 to 4.8°C in the highest emissions scenario. These findings have been recognised by the national science academies of the major industrial nations and are not disputed by any scientific body of national or international standing. The latest CO_2 emission data published late 2017 suggest that the figure has not plateaued as expected but continued to rise.

Figure 2.2 illustrates the change in global surface temperature relative to 1951–1980 average temperatures. Sixteen of the seventeen warmest years in the 136-year record are all since 2001, except for 1998. The year 2016 ranks as the warmest on record. This NASA/GISS[2] research is consistent with similar studies by other eminent organisations.

Future climate change and associated impacts will differ from region to region around the globe. Anticipated effects include increasing global temperatures, rising sea levels, changing precipitation, and expansion of deserts in the subtropics. Warming is expected to be greater over land than over the oceans and greatest in the Arctic, with the continuing retreat of glaciers, permafrost and sea ice. Other likely changes include more frequent extreme weather events.

Most countries are parties to the United Nations Framework Convention on Climate Change (UNFCCC), whose objective is to prevent dangerous anthropogenic climate change. Parties to the UNFCCC have agreed that deep cuts in emissions are required and that global warming should be limited to well below 2.0°C compared with pre-industrial levels, with efforts made to limit warming to 1.5°C.

Definition

Anthropogenic: Originating in human activity.

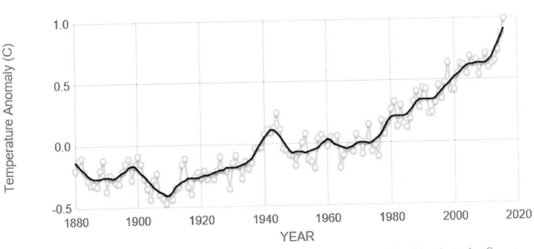

Figure 2.2 Global land-ocean temperature index. (Source: NASA's Goddard Institute for Space Studies (GISS))

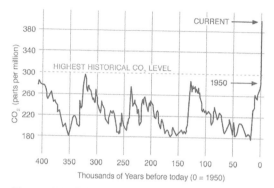

Figure 2.3 CO_2 levels reconstructed from ice cores. (Source: NASA/NOAA)

There is still a large gap between the pledges by governments to cut greenhouse gas emissions and the reductions scientists say are needed to avoid dangerous levels of climate change.

Current plans from national governments, and pledges made by private sector companies and local authorities across the world, could lead to temperature rises of as much as 3°C or more by the end of this century, far outstripping the goal set under the 2015 Paris agreement to hold warming to 2°C or less, which scientists say is the limit of safety.

There are signs that the world is moving away from its high-emissions trajectory. For instance, growing investment in renewable energy has caused the price of low-carbon power to plunge around the world, making it more attractive as an alternative to fossil fuels.

Key fact

Investment in renewable energy has caused the price of low-carbon power to fall.

The report also noted several potential short-term wins in staving off the worst effects of climate change, such as reducing the amount of soot entering the atmosphere, and phasing out the production of hydrofluorocarbons, used in air conditioning and refrigeration. The latter are powerful greenhouse gases, which, when they reach the atmosphere, cause warming many times greater than carbon dioxide, but there are alternatives that can be used instead. The UN also reported (November 2017) that atmospheric concentrations of carbon dioxide had reached record levels, in part owing to a strong El Niño weather system.[3]

Tables 2.2 and 2.3 present some figures to show CO_2 emissions and world population.

Table 2.4, on a slightly more positive note, shows how the production of biofuels has

Table 2.2 CO_2 emissions. (Source: BP Statistical Review of World Energy 2017)

	2006	2007	2008	2009	2010	2011	2012	2013	2014	2015	2016
Million tonnes of carbon dioxide	29430	30482	30800	30145	31528	32413	32740	33226	33343	33304	33432

Table 2.3 World population. (Source: www.worldometers.info)

	2006	2007	2008	2009	2010	2011	2012	2013	2014	2015	2016
World population (billion)	6.632	6.706	6.790	6.874	6.958	7.043	7.128	7.213	7.298	7.383	7.552

Table 2.4 Biofuels production. (Source: BP Statistical Review of World Energy 2017)

	2006	2007	2008	2009	2010	2011	2012	2013	2014	2015	2016
Biofuels production	27848	37471	50138	55936	64008	65834	66863	72293	79703	80024	82306

increased significantly over the ten years from 2006 to 2016. Further use of alternative fuels will help to reduce CO_2 (and other greenhouse emissions) but it seems there is much work to be done!

2.4 The hydrogen solution

A study entitled Hydrogen, scaling up[4] outlines a comprehensive and quantified roadmap to scale deployment and its enabling impact on the energy transition. It was presented to world leaders at COP23.[5] Deployed at scale, hydrogen could account for almost one-fifth of total final energy consumed by 2050. This would reduce annual CO_2 emissions by roughly 6 gigatons compared with today's levels, and contribute roughly 20% of the reduction required to limit global warming to 2°C.

On the demand side, the Hydrogen Council sees the potential for hydrogen to power about 10 to 15 million cars and 500,000 trucks by 2030, with many uses in other sectors as well, such as industry processes and feedstocks, building heating and power, power generation and storage. Overall, the study predicts that the annual demand for hydrogen could increase tenfold by 2050 to almost 80 EJ[6] in 2050, meeting 18% of total final energy demand in the 2050 2°C scenario. At a time when global population is expected to grow by 2 billion people by 2050, hydrogen technologies could have the potential to create opportunities for sustainable economic growth. Achieving such scale would require substantial investments within the right regulatory framework. The report considers that attracting these investments to scale the technology is feasible.

> **Definition**
> **Hydrogen Council:** Senior industry groups and representatives.

Hydrogen is a versatile, clean, and safe energy carrier that can be used as fuel for power or in industry as feedstock. Generating zero emissions at point of use, it can be produced from (renewable) electricity and from carbon-abated fossil fuels, thereby achieving completely zero-emission pathways. The uses for hydrogen continue to grow as it can be stored and transported at high energy density in liquid or gaseous form and can be combusted or used in fuel cells to generate heat and electricity.

2.5 Legislation and climate agreements

The Paris Agreement (French: Accord de Paris), Paris climate accord or Paris climate agreement, is an agreement within the United Nations Framework Convention on Climate Change (UNFCCC) dealing with greenhouse gas emissions mitigation, adaptation and finance starting in the year 2020. The language of the agreement was negotiated by representatives of 196 parties at the 21st Conference of the Parties of the UNFCCC in Paris and adopted by consensus on 12 December 2015. As of October 2017, 195 UNFCCC members have signed the agreement, and 169 have become party to it. The Agreement aims to respond to the global climate change threat by keeping a global temperature rise this century well below 2°C above pre-industrial levels and to pursue efforts to limit the temperature increase even further to 1.5°C.

> **Key fact**
> The Paris climate agreement was adopted by almost universal consensus on 12 December 2015.

In June 2017, US President Donald Trump announced his intention to withdraw the

United States from the Paris agreement, causing widespread condemnation in the European Union and many sectors in the United States. Under the agreement, the earliest effective date of withdrawal for the US is November 2020.

In July 2017, France's environment minister announced France's 5-year plan to ban all petrol and diesel vehicles by 2040 as part of the Paris Agreement. He also stated that France would no longer use coal to produce electricity after 2022 and that up to €4bn will be invested in boosting energy efficiency.

Other countries are starting to make similar declarations; the UK, for example, is to ban all new petrol and diesel cars and vans from 2040 amid fears that rising levels of nitrogen oxide pose a major risk to public health.

2.6 Real driving emissions

A new and improved method of testing vehicles for their real-world emissions has been developed. The test takes place on real roads and complements lab tests by measuring that a car delivers low pollutant emissions on the road.

The real driving emission (RDE) tests will measure the pollutants, such as nitrogen oxides (NOx), emitted by cars while driven on the road. RDE will not replace laboratory tests, such as the current NEDC and the future WLTP but it will be additional to them. RDE will ensure that cars deliver low emissions over on-road conditions. Europe will be the first region in the world to introduce such on-the-road testing, marking a major leap in the testing of car emissions.

Definition

RDE: Real driving emission.

Under RDE, a car will be driven on public roads and over a wide range of different

conditions. Specific equipment installed on the vehicle will collect data to verify that legislative caps for pollutants such as NOx are not exceeded. The test conditions include:

- low and high altitudes
- year-round temperatures
- additional vehicle payload
- up- and down-hill driving
- urban roads (low speed)
- rural roads (medium speed)
- motorways (high speed).

Definition

NOx: Nitrogen oxides

To measure pollutant emissions as the vehicle is being driven on the roads, cars will be fitted with Portable Emission Measuring Systems (PEMS) that will provide a complete real-time monitoring of the key pollutants emitted by the vehicle. The PEMS are complex pieces of equipment that integrate advanced gas analysers, exhaust mass flow meters, weather stations, GPS and a connection to the vehicle network.

RDE will almost certainly mean that diesel vehicles must be fitted with selective catalytic reduction (SCR) systems and some maybe also lean-NOx systems. This will mean additional costs for car manufacturers and smaller cars may not be able to accommodate

Figure 2.4 Equipment on a car for RDE testing

the fitting of SCR equipment. Prospective owners may be turned away by the additional costs.[7]

2.7 Euro emissions standards

2.7.1 Overview

European emission standards define the acceptable limits for exhaust emissions of new vehicles sold in EU and EEA member states. The emission standards are defined in a series of European Union directives detailing increasingly stringent standards. The stages are referred to as Euro 1, Euro 2, Euro 3, Euro 4, Euro 5 and Euro 6 for light vehicle standards. For heavy vehicles they are referred to as Euro I, Euro II, Euro III, Euro IV, Euro V, Euro VI.

The legal framework is a series of directives, each an amendment to the 1970 Directive (70/220/EEC). The following is a list of when the standards became mandatory for new vehicle registrations, together with the directive references:

▶ Euro 1 (1993): For passenger cars 91/441/EEC, also for passenger cars and light trucks 93/59/EEC.
▶ Euro 2 (1996) for passenger cars 94/12/EC (& 96/69/EC), for motorcycle – 2002/51/EC 2006/120/EC.

▶ Euro 3 (2000) for any vehicle 98/69/EC, for motorcycle 2002/51/EC and 2006/120/EC.
▶ Euro 4 (2005) for any vehicle 98/69/EC and 2002/80/EC.
▶ Euro 5 (2009) for light passenger and commercial vehicles 715/2007/EC.
▶ Euro 6 (2014) for light passenger and commercial vehicles 459/2012/EC.

2.7.2 Euro 6 for passenger cars

Euro 6 imposed a significant reduction in NOx emissions from diesel engines (67% reduction compared with Euro 5) and established similar standards for petrol and diesel. Exhaust gas recirculation (EGR) reduces the amount of nitrogen available to be oxidised to NOx during combustion but further exhaust after treatment may be required in addition to the diesel particulate filters (DPFs) required to meet Euro 5. To comply with Euro 6, diesel cars may also be fitted with:

▶ A NOx absorber (also known as a lean NOx trap), which stores NOx and reduces it to nitrogen by catalytic action.
▶ Selective catalytic reduction (SCR) using an additive (diesel exhaust fluid (DEF) or AdBlue) containing urea, which is injected into the exhaust to convert NOx into nitrogen and water.

> **Definition**
>
> **Urea:** Also known as carbamide, urea is an organic compound with the chemical formula $CO(NH_2)_2$ and when injected into the exhaust helps to convert NOx into nitrogen and water.

▶ The use of Cerium,[8] a fluid injected into the fuel tank each time the vehicle is refuelled which assists the DPF regeneration by nanoparticle catalytic action lowering the regeneration temperature.

Figure 2.5 It is expected that following Brexit, the UK will still comply with all the main EU directives

Euro 6 emission limits (petrol engines)

- ▶ CO – 1.0 g/km
- ▶ HC – 0.10 g/km
- ▶ NOx – 0.06 g/km
- ▶ PM – 0.005 g/km (direct injection only)
- ▶ PM – 6.0×10^{11}/km (direct injection only)

Euro 6 emission limits (diesel engines)

- ▶ CO – 0.50 g/km
- ▶ HC + NOx – 0.17 g/km
- ▶ NOx – 0.08 g/km
- ▶ PM – 0.005 g/km
- ▶ PM – 6.0×10^{11}/km

2.7.3 Euro VI emission standards for trucks and buses

The emission standards for vehicles for trucks (lorries) and buses are defined by engine energy output in g/kWh. Standards for passenger cars and light commercial vehicles are defined by vehicle driving distance in g/km. A direct comparison with passenger cars is therefore not possible because the kWh/km factor depends on many other issues relating to the specific vehicle under test. The official category name is heavy-duty diesel engines, which generally includes lorries and buses.

Euro 6 emission limits heavy-duty diesel engines

- ▶ CO – 1.50 g/kWh
- ▶ HC – 0.13 g/kWh
- ▶ NOx – 0.04 g/kWh
- ▶ PM – 0.01 g/kWh
- ▶ 8.0×10^{11} 1/kWh

The above results must be achieved during the world harmonised test cycle (WHTC). This test is a transient engine dynamometer schedule defined by the global technical regulation (GTR) No. 4, and developed by the UN ECE GRPE group. The regulation is based on the world-wide pattern of real heavy commercial vehicle use.

> **Definition**
> **WHTC:** World harmonised test cycle

Two representative test cycles, a transient test cycle (WHTC) with both cold and hot start requirements and a hot start steady-state test cycle (WHSC), have been created covering typical driving conditions in the EU, USA, Japan and Australia.

> **Definition**
> **WHSC:** World hot start steady-state test cycle.

The WHTC is a transient test of 1800s duration, with several motoring segments. Normalised engine speed and load (torque) values over the cycle are represented by Figure 2.6.

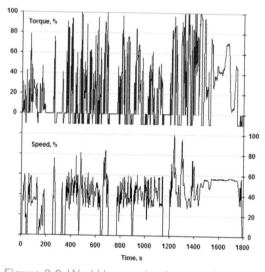

Figure 2.6 World harmonised test cycle

2.8 Summary

It can become somewhat depressing reading about climate change and associated catastrophes. When a major country pulls out of the Paris climate agreement it gets worse. When another country chooses to leave behind the good work being done in the EU, some of us feel even more depressed (some don't)!

It is none-the-less clear to anybody examining the evidence that something needs to be done to stop, or at least limit, global warming.

So, on a more positive note, because the use of alternative fuels and other technologies really can make a difference, read on and relax!

Notes

1 More information from the ACEA website: http://www.acea.be

2 https://climate.nasa.gov/vital-signs/global-temperature/

3 http://www.unenvironment.org/emissionsgap

4 http://hydrogencouncil.com/wp-content/uploads/2017/11/Hydrogen-scaling-up-Hydrogen-Council.pdf

5 COP23 is the informal name for the 23rd Conference of the Parties to the United Nations Framework Convention on Climate Change (UNFCCC). The UNFCCC established a framework for action to stabilise concentrations of greenhouse gases in the earth's atmosphere. The UNFCCC entered into force in 1994, and nearly all the world's nations (a total of 195) have now signed on. Each year the parties to the agreement convene to assess progress in implementing the convention and, more broadly, dealing with climate change.

6 EJ = 10^{18} joules.

7 More information here: http://www.caremissionstestingfacts.eu/rde-real-driving-emissions-test/

8 Catalytic Nanoparticles of cerium oxide are used.

CHAPTER 3

Alternative fuels

3.1 Introduction

There are arguably dozens of alternative fuels that can be used in some way to propel a vehicle. The use of an alternative fuel can lessen dependence upon oil and reduce greenhouse gas and other emissions. Alternative fuels can be described as regenerative if they are created by renewable processes such as biomass, wind or solar power. Hydrogen, which is usually created by electrolysis, can be considered regenerative providing a renewable electricity source is used.

> **Key fact**
>
> An alternative fuel can be described as regenerative if it is created by a renewable process.

All fuels except hydrogen release CO_2 during combustion. However, if fuels are derived from plants, the CO_2 created can be offset by that which is absorbed as the plants grow. There are several alternative fuels in common use and each of these is examined in this section.

> **Key fact**
>
> All fuels except hydrogen release CO_2 during combustion.

1 Ethanol
2 Methanol
3 Biodiesel
4 Natural gas
5 Liquified petroleum gas (LPG)
6 Hydrogen
7 Solar

A few 'new' fuels are being developed and some of these will also be outlined briefly.

Bio-fuels require a feedstock and currently many variations are under consideration or research for potential use. For example:

> **Definition**
>
> **Feedstock:** The raw material used to supply or fuel a machine or process.

▶ Soybean
▶ Corn
▶ Sugarcane
▶ Sugar beet
▶ Switchgrass

Alternative Fuel Vehicles. 978-1-138-50370-0 © 2018 Tom Denton.
Published by Taylor & Francis. All rights reserved.

3 Alternative fuels

- Jatropah (a flowering plant)
- Camelina (a flowering plant)
- Algae
- Cassava (a woody shrub)
- Palm oil
- Certain fungi
- Animal fat

3.2 Hydrocarbons

Petrol/gasoline and diesel are the fuels in common use. Petrol is known as an aliphatic hydrocarbon, which means it is made up of molecules that only contain hydrogen (H) and carbon (C). These molecules are arranged in chains and have from 7 to 10 carbon atoms in each chain. When petrol is burned under ideal conditions the result is carbon dioxide (CO_2) from the carbon and water (H_2O) from the hydrogen.

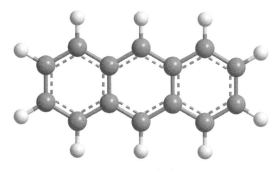

Figure 3.1 Hydrocarbon chain

chain and is known as methane. This gas is lighter than air but as the chains get longer the molecules get heavier. The first four chains, methane (CH_4), ethane (C_2H_6), propane (C_3H_8) and butane (C_4H_{10}) are all gases. The chains up through $C_{18}H_{32}$ are liquids at room temperature and above C_{19} they are semi-solids, fats and greases for example, and solids such as wax or bitumen.

> **Definition**
>
> **Aliphatic:** Organic compounds in which carbon atoms form open chains not aromatic rings.

> **Definition**
>
> **Methane (CH_4):** The lightest chain with just one carbon molecule.

Petrol contains 9.7kWh/L (thousand watt-hours per litre) – a lot of energy! To get an idea of this, 5L of petrol contains about the same energy as would be given off from a 1.5kW electric fire that is left on for 24 hours. If we could eat petrol, 5 litres would contain about 35,000 calories or the equivalent of 120 cheeseburgers!

> **Key fact**
>
> Petrol contains 9.7kWh/L of energy.

Petrol, and other hydrocarbons, are made from crude oil. These hydrocarbons have molecule chains with different lengths and they each have different properties. A chain with just one carbon molecule (CH_4) is the lightest

Because the different length chains all have progressively higher boiling points, they can be extracted from the crude oil by distillation in a refinery. As the oil is heated, the different chains are extracted at their vaporisation temperatures.

The chains C_5, C_6 and C_7 range are all light, clear liquids called naphthas. They are easily vaporised and are usually used as solvents and cleaning fluids. The chains from C_7H_{16} to $C_{11}H_{24}$ are blended together to form petrol/gasoline. They all vaporise below room temperature, which is why petrol evaporates if not in a sealed container. Kerosene is in the C_{12} to C_{15} range and this is followed by diesel and heating oils. After this come various thicknesses of lubrication oils and then semi-solid greases such as Vaseline. Finally, chains above C_{20} are solids ranging from paraffin wax, through tars to bitumen, which is used for making roads (when mixed with an aggregate).

Definition

Naphtha: Flammable oil containing hydrocarbons, distilled from organic substances such as coal, shale or petroleum.

The octane rating of petrol is an indication of how much it can be compressed before it ignites spontaneously. The higher the octane rating, also described using the research octane number (RON), the higher the compression ratio of the engine can be, and therefore more efficient and powerful. The number tends to range from 87 to 91. It is related to the percentage of octane, so a RON of 87 can have 87% octane and 13% heptane. However, the final number is determined by other additives as well as the amount of octane. For example, tetraethyl lead was used in the past to increase the octane number.

Diesel has a low octane number but a higher cetane value so is ideal for compression ignition engines. All sorts of additives have been used over many years, some with more success than others and some that turned out to be dangerous to health!

Key fact

Diesel has a low octane number but a high cetane value.

The moral of this long hydrocarbon-chain story is that as alternative fuels are created and mixed with existing fuels, it is essential that the resulting properties meet the requirements of the engines that they are to be used in.

3.3 Ethanol

Ethanol is a bio-alcohol-based fuel made by fermenting and distilling sugar or starch crops, such as corn or sugar cane. It can also be made from plants such as trees and grasses. It is made by first treating the plant with enzymes to split the long chain molecules into sugars such as dextrose. The next process involves fermenting the sugar with yeast to produce the ethanol. This method is like making spirits, wine and beer – but do not drink ethanol! The use of some types of enzyme means that the whole plant can be converted, not just the starchy components. However, there are some technical issues with this process.

Definition

Enzymes: Substances produced by living organisms, which act as a catalyst to cause a specific biochemical reaction.

Ethanol or bio-ethanol, is very suitable for mixing with petrol/gasoline and in doing this it also increases the octane rating. It can be used in a pure form in spark ignition flexible fuel vehicles (FFVs). However, because it can cause cold starting problems at low temperatures, it is used to a maximum of 85% (E85) in summer and 70–75% in winter. E10 is a blend of 10% ethanol and 90% petrol/gasoline. Almost all manufacturers approve the use of E10 in their vehicles. E95 is used by some petrol/gasoline powered heavy vehicles. Something in the region of 1.5 billion gallons (almost 7 million litres) of ethanol are blended with petrol/gasoline each year.[1] There are now over 1,000 stations offering E15 in the USA.

Definition

FFV: Flexible fuel vehicle.

Figure 3.2 E85 Ethanol pump in the USA. (Source: https://www.flickr.com/photos/diaper)

There is no noticeable difference in vehicle performance when E85 is used. However, FFVs operating on E85 can experience a 20–30% drop in miles per gallon due to ethanol's lower energy content. In low temperatures a minimum of a 70% mix is recommended.

There are some advantages and disadvantages of using ethanol:

Advantages

▶ Lower CO emissions
▶ Lower (25%) urban ozone-forming emissions
▶ More resistant to engine knock
▶ Normal fuelling systems can be used
▶ Added vehicle cost is small.

Disadvantages

▶ High percentages can only be used in flexible fuel

vehicles
▶ Lower energy content, resulting in reduced mileage (around 20%)
▶ Limited availability

3.4 Methanol

Methanol is not considered a regenerative fuel as it is usually produced from fossil sources such as natural gas and coal. It does not therefore, contribute to the reduction of CO_2. Some countries are using coal as a source of methanol to cover their high fuel demand. A 15% mixture with conventional fuel seems to be the upper limit (M15). A version of M85 is under investigation but it is more normal to use a 5% mixture (M5). Methanol can have a high corrosive effect on ferrous alloys so is problematic. It can also be used as a feeder for fuel cells, as discussed in Chapter 4.

Key fact
Methanol can have a high corrosive effect on steel.

Listed here are some advantages and disadvantages of methanol:

Advantages

▶ Can be used to meet high demand
▶ Can be used in some types of fuel cell.

Disadvantages

▶ It is not regenerative
▶ Corrosive
▶ Generally limited to a 5% mix

3.5 Biodiesel

Biodiesel is a form of diesel fuel manufactured from vegetable oils, animal fats or recycled restaurant oils. It is safe, biodegradable, and produces less air pollution than petroleum-based diesel. It is the most important alternative fuel for diesel engines at the time

Figure 3.3 Methanol dragster at Santa Pod. (Source: www.flickr.com/photos/sugarmonster)

of writing.

The term biodiesel covers fatty acid esters that are created through transesterification of oils and greases using ethanol or methanol. This process creates fatty acid methyl ester (FAME) or fatty acid ethyl ester (FAEE). Biodiesel molecules are very different from the vegetable oils biodiesel is made from and are very similar to diesel molecules. However, biodiesel can vary from petroleum diesel sometimes because the fatty acid esters are chemically reactive, unlike diesel which is very stable. The properties of the final product therefore depend on the feedstock as well as the process. There are various standards that can be used to control the quality.

> **Definitions**
>
> **FAME:** fatty acid methyl ester.
> **FAEE:** Fatty acid ethyl ester.

> **Key fact**
>
> Fatty acid esters are much more chemically reactive than diesel.

The feedstock for biodiesel is primarily rape seed oil in Europe, soya bean oil in North and South America, and in Asia it is palm oil. Used cooking oil can also be converted and is known as 'used frying oil methyl ester' (UFOME). The final products can be a mixture of the above sources.

One issue with biodiesel is that because methanol is the easiest substance to use for the esterification process, the result is not strictly regenerative as the methanol comes from coal. If ethanol is used, then the result is fully regenerative.

It can be used in its pure form (B100) but more strict emissions regulations and the use of diesel particulate filters (DPFs) mean that this is not common. It is therefore blended

Figure 3.4 Rape seed field

with petroleum diesel at lower percentages to ensure stability. A common blend is B7 (7%), which can be used safely in most diesel engines. However, many vehicle manufacturers do not recommend using blends greater than B5, and engine damage caused by higher blends is not covered by some manufacturer warranties. None-the-less, B20 is used by some diesel engines without any ill effects.

Biodiesel requires special attention when being stored and used in cold conditions. This is because it can freeze or gel. In very cold conditions therefore low cloud point additives may be needed. The cloud point is the lowest temperature at which a fuel can operate. It would normally be stored about 5°C above the cloud point.

> **Key fact**
> Biodiesel can freeze or gel when used in cold conditions.

On first use the cleaning effect of biodiesel can cause filters to clog but this is easily rectified with more frequent servicing in the first few weeks of use.

Listed here are some advantages and disadvantages of biodiesel:

Advantages

- ▶ Can be used in *many* diesel engines, especially newer ones at up to B20.
- ▶ Can be used in pure form in suitably modified engines.
- ▶ Fewer air pollutants (other than NOx) and fewer greenhouse gases.
- ▶ Biodegradable and non-toxic.
- ▶ Safer to handle.

- Lower flash point than pure diesel so it is considered safer.
- Normal dispensing equipment can be used.

Disadvantages

- Use of some blends may not be approved by manufacturers.
- B20 is 8% less energy dense than pure diesel.
- Lower fuel economy and power.
- More nitrogen oxide emissions.
- Concerns about impact on engine durability when higher concentrations are used.

3.6 Natural gas

Natural gas is a fossil fuel made up mostly of methane (83–98%), with the remainder being mostly inert gasses. It is one of the cleanest burning alternative fuels. The gas is available worldwide, but this introduces variations in density, calorific value and knock rating. This form of gas is not regenerative. However, another source of natural gas is biogas, which is made from energy crops, manure or waste biomass. This type is regenerative.

> **Definition**
> **Natural gas:** A fossil fuel made up mostly of methane.

Natural gas is stored either in gas form or as compressed natural gas (CNG). The pressure of this is about 200 bar. It can also be stored as a liquefied natural gas (LNG) at −162°C. LNG requires only one third of the volume of CNG; however, this does require significant energy to be used to liquefy it. For this reason, the most common form sold in fuel stations is compressed natural gas (CNG).

The amount of carbon in natural gas is lower than in some other fuels, so it produces much less CO_2 and more H_2O during combustion.

> **Definition**
> **Bar:** a unit of pressure defined as 100 kilopascals (kPa) approximately equal to the atmospheric pressure on Earth at sea level.

Dedicated natural gas vehicles are designed to run on natural gas only and spark ignition engines are usually used. Diesel engines can be used but in this case the gas is used as part of the pre-injection process. Dual-fuel vehicles take advantage of the widespread availability of conventional fuels and use the cleaner, more economical CNG when available. Natural gas is stored in high-pressure fuel tanks so dual-fuel vehicles require two separate fuelling systems, which take up extra space. These vehicles are not produced commercially in large numbers.

Some advantages and disadvantages of natural gas are noted here:

Advantages

- Fewer smog-producing particulates.
- 25% fewer greenhouse gas emissions.
- Huge reduction in urban ozone-forming emissions.
- Less expensive than petroleum fuels.

Figure 3.5 Honda CNG car. (Source: Mario Ortiz, https://creativecommons.org)

Disadvantages

▷ Limited vehicle availability.
▷ Less readily available.
▷ Disperses quickly if spilt.
▷ Fewer miles on a tank of fuel.

3.7 Liquefied petroleum gas (LPG)

Liquefied petroleum gas (LPG) is a clean-burning fossil fuel that can be used to power internal combustion engines. It is a mixture of propane and butane – and is often described as propane because that makes up most of the gas. It is refined from crude oil and it can be liquefied at room temperature under comparatively low pressure. Its low carbon content means it produces about 10% less CO_2 than the equivalent petrol/gasoline engine.

> **Definition**
>
> **LPG**: Liquid petroleum gas; a mixture of propane and butane.

Petrol/gasoline and diesel vehicles can be retrofitted to run on LPG in addition to conventional fuel. The LPG is stored in pressurised fuel tanks, so separate fuel systems are needed in vehicles powered by both LPG and a conventional fuel. It is known as Autogas in Germany, GPL in France and GLP in Italy and Spain.

Traditionally LPG vehicles used a vapour pressure system. Newer liquid injection systems have improved the efficiency considerably.

Following are advantages and disadvantages of LPG:

Advantages

▷ Fewer toxic and smog-forming particulates.
▷ Higher octane rating allowing for higher compression ratio and greater efficiency.
▷ Longest driving range compared with other alternative fuels.
▷ Spark plug life can be extended.
▷ Cheaper than petrol/gasoline.

Disadvantages

▷ Few cars and limited trucks commercially.
▷ Less readily available than conventional fuels.
▷ Special delivery and storage facilities are needed.

3.8 Hydrogen (H_2)

Hydrogen (H_2) can be produced from fossil fuels (such as coal or natural gas). Electricity from nuclear power, or renewable resources such as hydropower, can be used to create hydrogen by electrolysis of water. It is predominantly obtained by steam reformation from natural gas, but this method means that overall the CO_2 produced is about the same as conventional petrol/gasoline or diesel engines. It is cleaner at the point of use. Electrolysis from renewable sources is becoming more common.

> **Definition**
>
> **Hydrogen**: Colourless, odourless, highly flammable gas, chemical element with the atomic number 1.

Figure 3.6 Filling with LPG. (Source: www. flickr.com/photos/davilla)

It can be used in fuel cells to power electric motors or burned directly in internal combustion engines. Fuel cells are the much-preferred option as they are more efficient; these are examined later. Hydrogen is an environmentally friendly fuel at the point of use, as it only produces water as a by-product.

> **Definition**
>
> **Electrolysis:** Decomposition of a chemical by passing an electric current through a liquid or solution containing ions.

Hydrogen has a very high energy density (see Table 3.1) that is almost three times that of petrol/gasoline. However, it does have to be compressed to between 350 and 700 bar or liquefied at −253°C to achieve a suitable tank volume for a normal vehicle.

There are advantages and disadvantages to using hydrogen:

Advantages

▶ Can be produced from several sources, reducing dependence on petroleum.
▶ No air pollutants or greenhouse gases.
▶ It produces only NOx when burned in internal combustion engines.

Disadvantages

▶ Expensive to produce and is only available at a few locations.
▶ Fuel cell vehicles are currently too expensive for most consumers.
▶ Very high storage and delivery pressures are required.
▶ Hydrogen has a lower energy density than conventional petroleum fuels unless highly compressed.

3.9 Solar power

Solar power or energy cannot be stored in its basic form, but it can be produced by solar or photovoltaic panels, and used to charge a battery or create other alternative fuels such as hydrogen.

> **Definition**
>
> **Photovoltaic:** Production of electric current at the junction of two substances when exposed to light.

3.9.1 Solar powered hydrogen station

Honda[2] are working on a solar hydrogen station that is designed as a single, integrated unit to fit in the user's garage. The innovative design means the size of the system is quite small. It can produce enough hydrogen (0.5 kg) in 8 hours for daily commuting in a fuel cell electric vehicle. This would usually be overnight but anytime the vehicle is connected.

Figure 3.7 Hydrogen filling station

Next Generation Solar Hydrogen Station

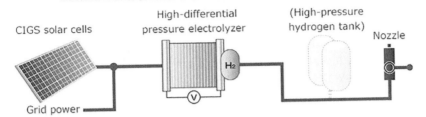

Current Solar Hydrogen Station

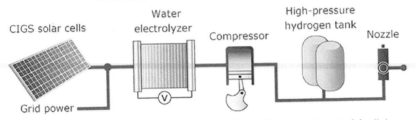

Figure 3.8 Current and next generation charging stations. (Source: Honda Media)

Key fact

Average daily commuting is considered to be around 10,000 miles per year (16,000 km).

The previous solar hydrogen station systems required both an electrolyser and a separate compressor unit to create high pressure hydrogen. The compressor was the largest and most expensive component and reduced system efficiency. By creating a new high differential pressure electrolyser, Honda engineers were able to eliminate the compressor entirely. This innovation also reduces the size of other key components.

Compatible with a 'Smart Grid' energy system, the Honda solar hydrogen station would enable users to refill their vehicle overnight without the requirement of hydrogen storage. This would lower CO_2 emissions by using less expensive off-peak electrical power. During daytime peak power times, the solar hydrogen station can export

Figure 3.9 Solar charging station. (Source: Honda Media)

Figure 3.10 Easy operation

Figure 3.11 Stella Vie (Source: TU Eindhoven, Bart van Overbeeke)

renewable electricity to the grid, providing a cost benefit to the customer, while remaining energy neutral.

As with the previous generation system, the hydrogen purity from the new station meets the highest SAE (J2719) and ISO (14687) specifications.

3.9.2 Stella Vie solar powered car

The Stella Vie[3] is a competition winning design for a solar powered car. A perfect balance between aerodynamic performance and aesthetic design was achieved. The result of this process is a car body that features some well-known automotive design features, whilst also being 9% more aerodynamic than its predecessor (Stella Lux).

The car can achieve a range of 1,000 km on a clear day. The surplus energy generated by Vie can be supplied back to the house or electric grid. A smart charging and discharging system keeps track of energy prices to find the optimal time to charge or discharge.

Through its updated solar navigator, which now takes height profile maps into account, Stella Vie finds the most efficient route and shows how much energy can be saved compared with a standard, fossil fuel-powered car. When approaching the destination, the car will offer parking assistance. If enabled, the built-in solar parking system will use height maps, weather data and a parking probability map to find a free parking space. The chosen parking spot is the one within range of the destination that yields the most solar energy.

To allow for safer and more efficient driving, Stella Vie uses the latest V2X technology to warn the driver to anticipate upcoming traffic events. It also encourages the user to drive as efficiently as possible by giving subtle feedback through a built-in lighting system; this system warns the driver by turning red when either braking or accelerating too fiercely.

3.10 Other fuels

3.10.1 Dimethyl ether

Dimethyl ether (DME) is a fuel under development for use in diesel engines, petrol/gasoline engines and gas turbines. For petrol/gasoline it is usually used as a 30% DME and 70% LPG mixture. DME has a cetane number of 55 and diesel is 40–53. Only moderate modifications are therefore needed to convert a diesel engine to burn DME.

Because of a simple chemical structure, combustion results in very low emissions of particulate matter, NOx, and CO. It is also sulfur-free so can meet stringent emission regulations.

Although lower than conventional cars, due to the big doors and handles, it is easy to enter. Inside, the car provides a comfortable driving experience due to ergonomically optimised seats, a clean and spacious interior and intuitive controls for both driving and infotainment. Stella Vie is the first solar powered family car to seat five people.

Figure 3.12 Stella Vie is charging in the sun while parked. (Source: TU Eindhoven, Bart van Overbeeke)

DME is primarily produced by converting natural gas, organic waste or biomass to synthesis gas (syngas). The syngas is then converted into DME via a two-step synthesis, first to methanol in the presence of a catalyst (usually copper-based), and then by subsequent methanol dehydration in the presence of a different catalyst (for example, silica-alumina) into DME.

3.10.2 P-Series

P-Series fuel was developed in the 1990s by thermonuclear physicist Stephen Paul of Princeton University (the P is for Princeton). It is a blend of 35% natural gas liquids, 45% ethanol and 20% methyltetrahydrofuan (MeTHF). The natural gas liquid is known as pentanes-plus, which is what is left over after the processing of natural gas.

MeTHF can be made from biomasses that have a negative cost such as paper sludge, food waste and agricultural waste. The feedstock used to make MeTHF is chemically digested by the process of making it and, as a result, the raw material is completely consumed. No emissions are produced during production, which means that burning P-Series fuels in vehicles releases fewer emissions than pure fossil fuels.

The fuel has an octane value between 89 and 93, which is the same as petrol/gasoline. It can be formulated for winter (by adding butane) or summer use, used on its own, or mixed with petrol/gasoline.

Flexible fuel vehicles (FFVs) designed to burn methanol or ethanol can use P-Series, but it is not suitable for ordinary cars. P-Series is about 10% less efficient in use than petrol/gasoline, but the advantage of recycling biomass waste and lower emissions make it an interesting alternative fuel.

3.10.3 Synthetic fuels

For climate targets to be achieved, CO_2 emissions from traffic worldwide will have to be reduced by 50% over the next four decades, and by at least 85% in the advanced economies. After all, even if all cars were to drive electrically one day, aircraft, ships, and most trucks will still run mainly on fuel. Carbon-neutral combustion engines that run on synthetic fuels[4] are thus a very promising path to explore.

Synthetic, or carbon-neutral, fuels capture CO_2 in the manufacturing process. In this way, this greenhouse gas becomes a raw material, from which gasoline, diesel and substitute natural gas can be produced with the help of electricity from renewable sources. One further crucial advantage of the combustion engine using synthetic fuels is that the existing filling-station network can continue to be used.

Definition
Carbon-neutral: Making or doing something that results in no *net* release of carbon dioxide into the atmosphere.

The same applies to the existing combustion-engine expertise. Even though electric cars will become significantly less expensive in the years ahead, the development of these fuels may still be worthwhile. Bosch has calculated that, up to a lifetime mileage of 160,000 km, the total cost of ownership of a hybrid running on synthetic fuel could be less than that of a long-range electric car, depending on the type of renewable energy used.

However, considerable efforts are still needed before synthetic fuels can become established. The processing facilities are still expensive, and there are only a few test plants. The widespread use of these fuels will also be helped by the increasing availability of, and thus falling prices for, electricity from renewables.

Synthetic fuels are made solely with the help of renewable energy. In a first stage, hydrogen is produced from water. Carbon is then added to produce a liquid fuel. This carbon can be recycled from industrial processes or even captured from the air using filters. Combining CO_2 and H_2 results in the synthetic fuel, which can be petrol/gasoline, diesel, gas or even kerosene.

Synthetic fuels do not mean a choice between fuel tank and dinner plate, as biofuels do. And if renewable energy is used, synthetic fuels can be produced without the volume limitations that can be expected in the case of biofuels because of factors such as the amount of land available. It is possible that by 2025 synthetic fuels will be available commercially.

3.10.4 Algae fuel

Algae fuel uses algae as its source of energy-rich oils. The algae are grown in pods or beds or transparent tubes. As a fuel, it only releases CO_2 recently removed from the atmosphere via photosynthesis as the algae or plant grew. The algae can be grown with minimal impact on fresh water resources and can even be produced using saline and wastewater. The resulting fuels have a high flash point and are biodegradable so are relatively harmless to the environment if spilled.

Definition

Algae: Simple, non-flowering, and usually aquatic plants that include seaweeds and many single-celled forms. Algae contain chlorophyll but lack stems, roots and leaves.

Figure 3.13 Algae growing in tubes to allow greater photosynthesis

Algae cost more per unit mass than other biofuel crops due to high capital and operating costs, but are claimed to yield between 10 and 100 times more fuel per unit area. It is estimated that if algae fuel replaced all the petroleum fuel in the US, it would require 15,000 square miles (39,000 km²), which is less than half a per cent of the US land area. Corn harvested in the US in 2000 used about 3.5% of the land area. Traditional refinery methods, to produce different grades of fuel, can be used on the raw materials produced by algae.

A recent news update stated that scientists from Synthetic Genomics (SGI) and ExxonMobil have developed a strain of fatty algae able to convert carbon into a record amount of energy-rich fat, which can then be processed into biodiesel. While algae using CO_2 to generate fat is not new, the amount of fat produced by the algae is noteworthy.[5]

Success in developing algae-based biodiesel at commercial levels will provide several benefits. The fuel emits fewer greenhouse gases than most conventional energy sources, and that will help strengthen our ongoing transition to low-emission energy resources.

Also, unlike other biofuel feedstocks, such as corn, algae production at an industrial scale would not impact on food production.

To boost fat production, SGI scientists worked with their counterparts at ExxonMobil as nutritionists of sorts, tweaking the part of the algae genome responsible for the assimilation of nitrogen, an essential nutrient. The change is algae with about 40% of its mass as fat. That's more than double the fat content of conventional algae.

Oil from algae can also potentially be processed in conventional refineries, producing fuels no different from convenient, energy-dense diesel. Oil produced from algae also holds promise as a potential feedstock for chemical manufacturing.

3.11 Properties of fuels compared

The information in Tables 3.1 and 3.2 is supplied to allow general comparisons to be made and to show some key properties.[6] Use the figures in Tables 3.1 and 3.2 as a guide and refer to other sources for specific details.

Table 3.1 Liquid fuel energy storage and properties

Fuel type	Specific energy (MJ/kg)	Density (kg/l)	A/F ratio (by mass)	Anti-knock index (AKI)[7]	Research octane number (RON)[8] or cetane
Petrol/Gasoline	42	0.75	14.7	85 to 96	90 to 105
Diesel	43	0.85	14.5		48–52 (cetane number)
Biodiesel	37	0.88	12.8		
Methanol	20	0.79	6.4	98	108
Ethanol	27	0.79	9.0	99	108
Liquefied natural gas	46	0.54	15.5		
Methane	56	0.66	17.2		
Propane	50	0.51	15.6	108	118
Butane	50	0.58	15.4	97	103
Hydrogen	120	0.07	34.0		
DME	28	0.67	14–15		55–60 (cetane number)
Uranium	80,000,000	19.1			

Table 3.2 Fuel cell and battery energy storage

Storage type	Specific energy (MJ/kg)
Direct-Methanol fuel cell	4.6
Proton-Exchange fuel cell	5.7
Lead-acid battery	0.2
Nickel metal hydride battery	0.3
Lithium ion battery	0.8

Notes

1 US Department of Energy.
2 Source: http://world.honda.com/worldnews/2010/c100127New-Solar-Hydrogen-Station.html
3 Stella Vie Team Eindhoven: https://solarteameindhoven.nl/stella-vie/
4 Bosch: www.bosch.com/explore-and-experience/synthetic-fuels
5 Synthetic Genomics: https://www.syntheticgenomics.com/exxonmobil-and-synthetic-genomics-report-breakthrough-in-algae-biofuel-research/
6 Sources for Tables 3.1 and 3.2: *Automotive Handbook*, 9th edition 2014. Robert Bosch GmbH. SAE. Wikipedia: https://en.wikipedia.org/wiki/Energy_density. Neutrium: https://neutrium.net/properties/specific-energy-and-energy-density-of-fuels. Biofuel: http://biofuel.org.uk
7 Used in the USA and Canada.
8 Used in Europe and Australia.

CHAPTER 4

Fuel cells

4.1 Fuel cells

4.1.1 Hydrogen fuel cells

Proton-exchange membrane fuel cells, also known as polymer electrolyte membrane (PEM) fuel cells (PEMFC), are the type mainly used for transport applications. They can operate at a lower temperature (50 to 100°C) than earlier types. They use a special proton-conducting polymer electrolyte membrane.

Fuel cells produce electricity through the reaction of a fuel with oxygen. Hydrogen-oxygen fuels cells use hydrogen as the fuel. Water is the only waste product from this type of fuel cell. The reaction between hydrogen and oxygen is exothermic, in other words it releases energy. The balanced chemical equation is:

▶ $2H_2 + O_2 \rightarrow 2H_2O$ (hydrogen + oxygen → water)

Definitions

Exothermic: A chemical reaction that transfers energy to the surroundings.

Endothermic: A chemical reaction that takes in energy from the surroundings.

Fuel cells use this reaction between hydrogen and oxygen to produce electrical energy. They convert a large proportion of the chemical energy into electrical energy so are very efficient.

All oxidations involve a transfer of electrons between the fuel and oxidant, and this is employed in a fuel cell to convert the energy directly into electricity.

All battery cells involve an oxide reduction at the positive pole and an oxidation at the negative during some part of their chemical process. To achieve the separation of these reactions in a fuel cell, an anode, a cathode, electrolyte and a membrane are used. The electrolyte is fed directly with the fuel (hydrogen).

Key fact

The energy of oxidation of conventional fuels can be converted directly into electricity in a fuel cell.

Fuel cells are very reliable and silent in operation, but can be quite expensive to construct. They are now usually described

4 Fuel cells

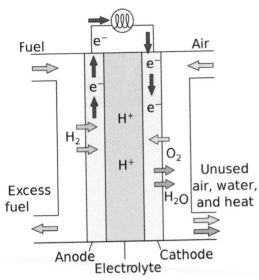

as proton exchange membrane fuel cells (PEMFC).

> **Key fact**
>
> The maximum theoretical energy efficiency of a fuel cell operating at low power density, using pure hydrogen and oxygen and without additional energy capture, is 83%. For an internal combustion engine, the figure is about 58%.

Operation of a hydrogen PEM fuel cell is as follows:

1 Hydrogen is passed over an electrode (the anode), which is coated with a catalyst.
2 The hydrogen (H) diffuses into the electrolyte membrane.
3 Electrons are stripped off the hydrogen atoms because the membrane only allows protons to pass.
4 The electrons then pass through the external circuit.
5 Negatively charged hydrogen anions (OH) are formed at the electrode over which

oxygen is passed such that it also diffuses into the solution.
6 These anions move through the electrolyte to the anode.
7 Water (H_2O) is formed as the by-product of a reaction involving the hydrogen ions, electrons and oxygen atoms.

If the heat generated by the fuel cell is used (energy capture), an efficiency of over 80% is possible, together with a very good energy density figure.

> **Key fact**
>
> A unit consisting of many individual fuel cells is referred to as a stack.

The working temperature of these cells varies but about 200°C is typical. High pressure is also used, and this can be of the region of 30 bar. The pressures and storage of hydrogen were the main problems, but these have now been overcome so the fuel cell can be a realistic alternative to other forms of storage.

> **Safety first**
>
> The working temperature of fuel cells varies but up to 200°C is possible. High pressure is also used, and this can be of the order of 30 bar.

For the two electrodes in a hydrogen-oxygen fuel cell, the balanced equations of the process are as follows:

▶ At the cathode (negative electrode):
  $H2(g) - 2e - \rightarrow 2H + (aq)$
▶ At the anode (positive electrode):
  $4H + (aq) + O2(g) + 4e - \rightarrow 2H2O(g)$

(g = gas, aq = aqueous)

The reaction at the cathode is an oxidation reaction because hydrogen loses electrons, and the reaction at the anode is a reduction reaction because hydrogen ions gain electrons. The overall reaction in the fuel cell is a redox reaction.

The cells of a fuel cell are connected in series just like a normal battery to form a stack that has the necessary voltage and current supply requirements. The theoretical open circuit voltage of a hydrogen-oxygen fuel cell is 1.23 V at room temperature; in practice it is around 1 V at open circuit. Under load conditions, the cell voltage is between 0.5 and 0.8 V. The stack voltage is usually 400–500 V.

Figure 4.2 Honda fuel cell stack. (Source: Honda Media)

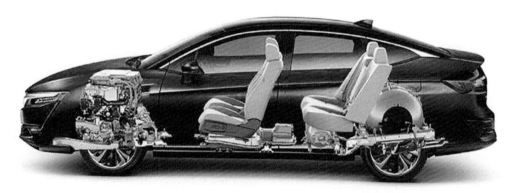

Figure 4.3 In the Honda Clarity (2017) the stack is incorporated with the motor under the bonnet. (Source: Honda Media)

37

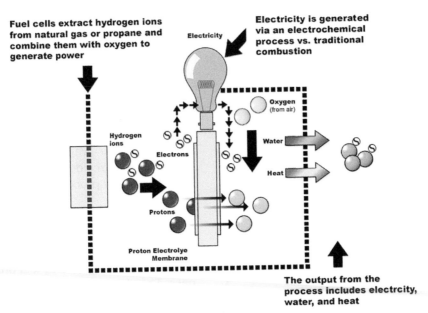

Fuel cells extract hydrogen ions from natural gas or propane and combine them with oxygen to generate power

Electricity is generated via an electrochemical process vs. traditional combustion

Electricity

Oxygen (from air)

Hydrogen ions

Electrons

Water

Heat

Protons

Proton Electrolye Membrane

The output from the process includes electrcity, water, and heat

Figure 4.4 Fuel cell operation. (Source: Dana)

4.1.2 Methanol fuel cells

Many combinations of fuel and oxidant can be used for fuel cells. Though hydrogen and oxygen as a combination is conceptually simple, hydrogen has some practical difficulties, including that it is a gas at standard temperature and pressure. In addition, an infrastructure for distributing hydrogen to domestic users does not exist. Fuel cells have therefore also been developed to run on methanol. There are two basic types of fuel cell that use this fuel:

▶ Reformed methanol fuel cell (RMFC)
▶ Direct methanol fuel cell (DMFC).

In the RMFC, a reaction is used to release hydrogen from the methanol, and then the fuel cell runs on hydrogen. The methanol is used as a carrier for hydrogen. The DMFC uses methanol directly. RMFCs can be made to be more efficient than DMFCs, but they are more complex.

DMFCs are a type of proton exchange membrane fuel cell (PEMFC). The membrane in a PEMFC fulfils the role of the electrolyte, and the protons (positively charged hydrogen ions) carry electrical charge between the electrodes.

Because the fuel in a DMFC is methanol, not hydrogen, other reactions take place at the anode. Methanol is a hydrocarbon (HC) fuel, which means that its molecules contain hydrogen and carbon (as well as oxygen in the case of methanol). When HCs burn, the hydrogen reacts with oxygen to create water and the carbon reacts with oxygen to create carbon dioxide. The same general process takes place in a DMFC, but in the process the hydrogen crosses the membrane as an ion, in just the same way as it does in a hydrogen-fuelled PEMFC.

> **Key fact**
> A benefit of methanol is that it can easily fit into the existing fuel infrastructure of filling stations; however, for automotive use the hydrogen fuel cell is proving to be the best option.

CHAPTER 5

Engines

5.1 Introduction

This chapter will be revision for many readers, but it will serve as a useful reminder of some technologies and why they are relevant to alternative fuels.

The key to the success of all modern engines and associated systems is very accurate control of their operation (fuelling and ignition, for example). Changing to a different fuel or even mixing in a small percentage of a different fuel by adding biodiesel to diesel, or ethanol to petrol/gasoline, can have a big effect on operation. For this reason, some 'dual fuel' vehicle have two separate fuel control systems. These may talk to each other to share sensor data such as engine speed but otherwise the operating data is very different.

> **Key fact**
>
> The secret to successful operation of all modern IC engines is accurate electronic control of fuelling, ignition and other engine management functions.

Engine and fuel system materials must also be modified to prevent corrosion by some fuels and the increased need for resistance to wear. For example, in addition to cleaner fuel combustion, natural gas has a higher anti-knock index compared with petrol. The octane number for natural gas can be up to 130 RON. This allows an earlier ignition time without combustion knock. The efficiency increases and the combustion pressure and combustion temperature in the combustion chamber also increase. However, natural gas is very dry and does not have lubricating properties like petrol. All these factors place additional strains on the engine and require modifications to the mechanical components.

Some typical mechanical modifications to the standard Volkswagen Golf 1.4, 81 kW TGI engine, so it can be run on petrol or natural gas, are listed in Table 5.1.

5.2 Engines and engine management

5.2.1 Four-stroke cycle

Figure 5.1 shows a modern vehicle engine. Engines like this can seem very complex at first but keep in mind that, with very few exceptions, all engines operate on the four-stroke principle. The complexity is in the systems around the engine to make it

Table 5.1 Component modifications. (Source: Volkswagen)

Pistons and rings	Aluminium pistons are hard anodised in the first ring groove and the uppermost piston ring has a special coating to give increased wear resistance.
Camshaft and valve timing	The closing segment of the inlet and exhaust cams are not as steep, so the valves are closed more slowly, and mechanical load is reduced.
Valves, valve guides, valve stem oil seals and valve seat inserts	The inlet and exhaust valves are nitrated, plated and have hardened shafts to make them more wear-resistant. The material used for the inlet valve guides and valve seat inserts has been adapted. The valve stem oil seals on the exhaust valves have two sealing lips with a reduced radial force to increase oil flow and improve the lubrication between the valve shafts and guides.
High-pressure injectors	In petrol mode the high-pressure injectors are cooled by the flowing fuel (petrol). Because there is no cooling in natural gas mode, and these valves protrude directly into the combustion chamber, very high temperatures would result. A graphitic Teflon ring with a high thermal conductivity is therefore used.
Turbocharger	The high efficiency of the natural gas operation further reduces the energy content in the exhaust gas. A smaller compressor wheel is therefor used to achieve quick response characteristics.

Figure 5.1 Volkswagen Golf 1.4 |81 kW TGI engine. (Source: Volkswagen)

operate to its maximum efficiency or best performance.

The engine components are combined to use the power of expanding gas to drive the engine. When the term 'stroke' is used it means the movement of a piston from top dead centre (TDC) to bottom dead centre (BDC) or the other way around. The following table explains the spark ignition (SI) and compression ignition (CI) four stroke cycles. Figure 5.2 shows the SI four-stroke cycle.

Definitions

TDC: Top dead centre.

BDC: Bottom dead centre.

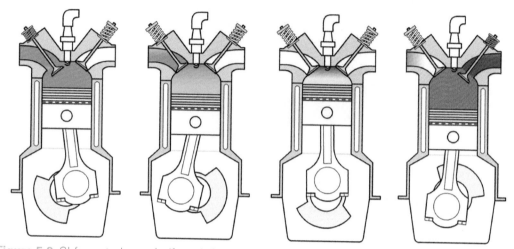

Figure 5.2 SI four-stroke cycle (from left to right, induction, compression, power, exhaust)

Table 5.2 Four-stroke cycle operation

Stroke	Spark ignition	Compression ignition
Induction	The fuel air mixture is forced into the cylinder through the open inlet valve because as the piston moves down it makes a lower pressure. It is acceptable to say the mixture is drawn into the cylinder.	Air is forced into the cylinder through the open inlet valve because as the piston moves down it makes a lower pressure. It is acceptable to say the air is drawn into the cylinder.
Compression	As the piston moves back up the cylinder the fuel air mixture is compressed to about an eighth of its original volume because the inlet and exhaust valves are closed. This is a compression ratio of 8:1, which is typical for many normal engines.	As the piston moves back up the cylinder the fuel air mixture is compressed in some engines to about a sixteenth of its original volume because the inlet and exhaust valves are closed. This is a compression ratio of 16:1, which causes a large build-up of heat.
Power	At a suitable time before top dead centre, a spark at the plug ignites the compressed mixture. The mixture now burns very quickly, and the powerful expansion pushes the piston back down the cylinder. Both valves are closed.	At a suitable time before top dead centre, very high pressure atomised diesel fuel (at about 180 bar), is injected into the combustion chamber. The mixture burns very quickly, and the powerful expansion pushes the piston back down the cylinder. The valves are closed.
Exhaust	The final stroke occurs as the piston moves back up the cylinder and pushes the spent gases out of the now open exhaust valve.	The final stroke occurs as the piston moves back up the cylinder and pushes the spent gases out of the now open exhaust valve.

Figure 5.3 Volkswagen 2.0 L, 176 kW TDI BiTurbo engine. (Source: Volkswagen Media)

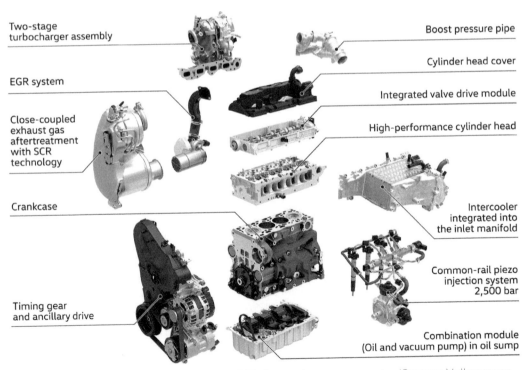

Figure 5.4 Volkswagen 2.0 L, 176 kW TDI BiTurbo engine components. (Source: Volkswagen Media)

5.2.2 Cylinder layouts

Improvements can be made to the performance and balance of an engine by using more than one cylinder. The layout can be one of three possibilities, as follows:

- ▶ In line or straight, where the cylinders are in a straight line; they can be vertical, inclined or horizontal.
- ▶ Vee, where the cylinders are in two rows at a set angle, which varies but is often 60° or 90°.
- ▶ Opposed, where the cylinders are in two rows opposing each other and are usually horizontal.

By far the most common arrangement is the straight four and this is used by all manufacturers in their standard family cars. Larger cars do however make use of the vee configuration. The opposed layout while still used is less popular.

Key fact

The most common cylinder arrangement is the straight four, but smaller three-cylinder engines are gaining popularity.

5.2.3 Camshaft drives

The engine drives the camshaft in one of three ways: gear drive, chain drive or by a drive belt. The last of these is now the most popular as it tends to be more simple and quieter. Note in all cases that the cam is driven at half the engine speed. This is done by the ratio of teeth between the crank and cam cogs, which is 1:2, for example, 20 crank teeth and 40 cam teeth.

Key fact

An engine camshaft is always driven at half the crankshaft (engine) speed.

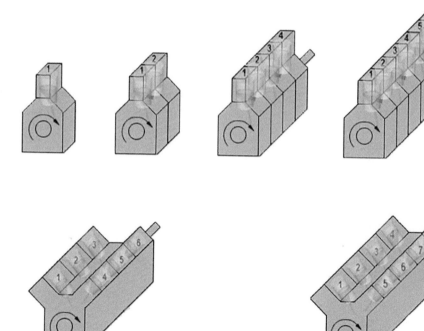

Figure 5.5 Common cylinder configurations

Figure 5.6 Variable camshaft timing. (Source: Ford Media)

▶ Camshaft drive gears are not used very often on petrol engines but are used on larger diesel engines. When they ensure a good positive drive from the crankshaft gear to the camshaft.

▶ Camshaft chain drive is still used but was even more popular a few years ago. The problems with it are that a way must be found to tension the chain and provide lubrication.

▶ Camshaft drive belts have become very popular. The main reasons for this are that they are quieter, do not need lubrication and are less complicated. They do break now and then but this is usually due to lack of servicing.

5.2.4 Valve mechanisms

Several methods are used to operate the valves. Three common types that use an overhead camshaft (as most engines now do) are shown as Figure 5.7, and a basic explanation of each follows (left to right):

▶ **Overhead cam with rockers**: As the cam turns it moves the follower, which in turn pushes the push valve open. As the cam move further it allows the spring to close the valve.

▶ **Overhead cam, with rocker and automatic adjusters**: Most new engines now use an OHC with automatic adjustment. This saves repair and service time and keeps the cost to the customer lower. Systems vary between manufacturers, some use followers and some have the cam acting directly on to the valve (Figure 5.8). In each case though the adjustment is by oil pressure. A type of plunger, which has a chamber where oil can be pumped under pressure, operates

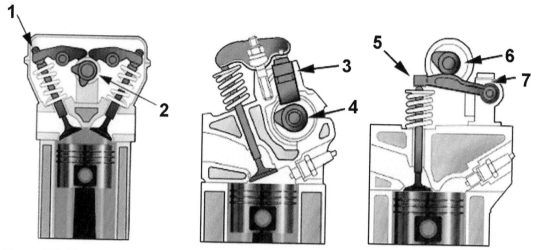

Figure 5.7 Alternative OHC valve operating mechanisms: 1. Valve, 2. Cam, 3. Hydraulic cam follower, 4. Cam, 5. Valve rocker, 6. Cam, 7. Pivot

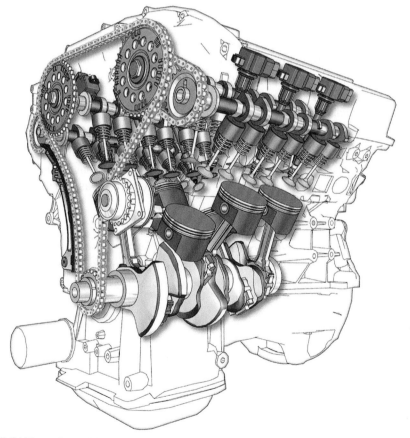

Figure 5.8 DOHC engine with direct acting camshafts and hydraulic followers. (Source: Swaroopvarma, Wikimedia Commons)

the valve. This expands the plunger and takes up any unwanted clearance.

▶ **Overhead cam with followers**: In the system shown here the lobe of the cam acts directly on the follower, which pivots on its adjuster and pushes the valve open.

5.2.5 Valve and ignition timing

Valve timing is important, Figure 5.9 is a diagram showing the degrees of rotation of the crankshaft where the inlet and exhaust valves open and close during the four-stroke cycle. The actual position in the cycle of operation when valves open and close depends on many factors and will vary slightly with different designs of engine. Many cars now control valve timing by electronics (Figure 5.6).

The valve timing diagram shows that the valves of a four-stroke engine open just before and close just after the related stroke. Looking at the timing diagram, if you start at position 2, the piston is nearly at the top of the exhaust stroke when the inlet valve opens (IVO). The piston reaches the top and then moves down on the intake stroke. Just after starting the compression stroke the inlet valve closes (IVC). The piston continues upwards and at a point several degrees before top dead centre; the spark occurs and starts the mixture burning.

The maximum expansion is 'timed' to occur after top dead centre, therefore the piston is pushed down on its power stroke. Before the end of this stroke (at position 5) the exhaust valve opens (EVO). Most of the exhaust gases now leave because of their very high pressure. The piston pushes the rest of the spent gases out, as it moves back up the cylinder. The exhaust valve closes (EVC), just after the end of this stroke and the inlet has already opened, ready to start the cycle once again. This makes the engine more efficient by giving more time for the mixture to enter and the spent gases to leave. The outgoing

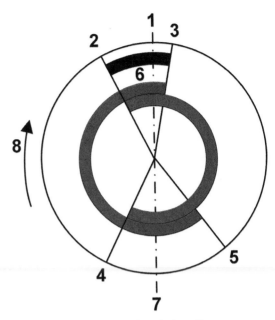

Figure 5.9 Circular valve timing diagram: 1. TDC, 2. Inlet valve opens, 3. Exhaust valve closes, 4. Inlet valve closes, 5. Exhaust valve opens, 6. Valve overlap, 7. BDC, 8. Rotation (two rotations for the full cycle)

exhaust gases in fact help to draw in the fuel air mixture from the inlet. Over all, this makes the engine have a better 'volumetric efficiency'. This phase (6) is known as valve overlap.

> **Key fact**
> The maximum expansion of a combusting fuel air mixture occurs after top dead centre.

5.3 Combustion

5.3.1 Introduction

The process of combustion in spark and compression ignition engines is best considered for petrol and diesel engines in turn. The combustion section here will give

Figure 5.10 Diesel combustion. (Source: Bosch Media)

enough information to allow considered opinion about the design and operation of electronic fuel control systems. In the context of alternative fuels, it also illustrates why any changes to the type of fuel used are quite complex.

A simplified description of the combustion process within the cylinder of spark ignition (SI) engine is as follows. A single high intensity spark of high temperature passes between the electrodes of the spark plug leaving behind it a thin thread of flame. From this thin thread, combustion spreads to the envelope of mixture immediately surrounding it at a rate that depends mainly on the flame front temperature, but also, to a lesser degree, on the temperature and density of the surrounding envelope.

In this way a bubble of flame is built up that spreads radially outwards until the whole mass of mixture is burning, the bubble containing the highly heated products of combustion, while ahead of it, and being compressed by it, lies the still unburnt mixture.

If the cylinder contents were at rest this bubble would be unbroken but with the turbulence and of air normally present within the cylinder the filament of flame is broken up into a ragged front, which increases its area and greatly increases the speed of advance. While the rate of advance depends on the degree of turbulence the direction is but not affected unless some definite swirl is imposed on the system. The combustion can be considered in two stages:

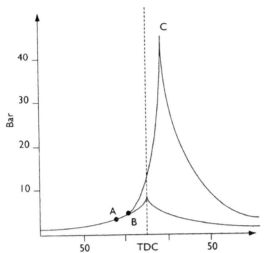

Figure 5.11 Cylinder pressure rise after ignition at point A and the delay to point B

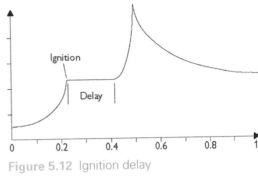

Figure 5.12 Ignition delay

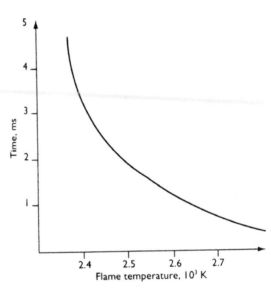

Figure 5.13 Approximate relation between flame temperature and the time from spark to propagation

1 Growth of a self-propagating flame.
2 Spread through the combustion chamber.

The first process is chemical and depends on the nature of the fuel, the temperature and pressure at the time, and the speed at which the fuel will oxidise or burn. Shown in Figure 5.11 it appears as the interval from the passage of the spark (A) to the time when an increase in pressure due to combustion can first be detected (B).

> **Key fact**
>
> Combustion can be considered in two stages: Growth of a self-propagating flame and then the spread through the combustion chamber.

If fuel is burned at a constant volume having been compressed to a self-ignition temperature, the pressure-time relationship appears as in Figure 5.12. The time interval occurs with all fuels but may be reduced with an increase of compression temperature.

With the combustion under way, returning to the cylinder pressure diagram Figure 5.11,

the pressure rises within the engine cylinder from (B) to (C) very rapidly approaching the 'constant volume' process of the four-stroke cycle. While (C) represents the peak cylinder pressure and the completion of flame travel, all available heat has not been liberated due to re-association and what can be referred to as after burning, which continues throughout the expansion stroke.

The graphs shown as Figures 5.13–5.15 show that the minimum delay time (A to B) is about 0.2 ms seconds with the mixture slightly rich. While the second stage (B to C) is roughly

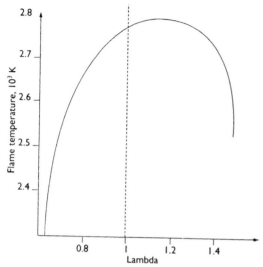

Figure 5.14 Relationship between flame temperature and the mixture strength

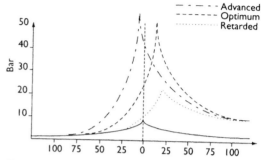

Figure 5.16 Effects of ignition timing

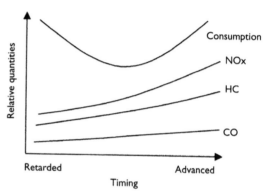

Figure 5.17 Effect of spark timing on consumption and emissions

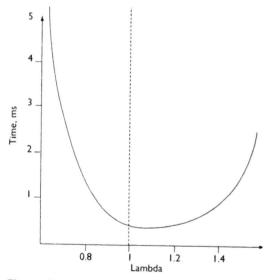

Figure 5.15 Relationship between mixture strength and rate of burning

dependent upon the degree of the turbulence, the initial delay necessitates ignition advance as the engine speed increases.

As ignition is advanced there is an increase in firing pressure or maximum cylinder pressure, generally accompanied by a reduction in

exhaust temperature. The effect of increasing the range of the mixture strength speeds the whole process up and thus increases the tendency to detonate.

Definition

Ignition advance: The change in timing so the spark comes earlier in the cycle

5.3.2 Detonation and pre-ignition

Detonation is the limiting factor on the output and efficiency of the spark ignition engine. The mechanism of detonation is the setting up within the engine cylinder of a pressure wave travelling at such velocity

49

as, by its impact against the cylinder walls, to set them in vibration and thus produce a high pitched 'ping'. When the spark ignites a combustible mixture of the fuel and air, a small nucleus of flame builds up, slowly at first but it accelerates rapidly. As the flame front advances, it compresses ahead of it the remaining unburned mixture. This raises the temperature of the unburned mixture by compression and radiation from the advancing flame until the remaining charge ignites spontaneously. The detonation pressure wave passes through the burning mixture at a very high velocity and the cylinder walls emit the ringing knock.

High compression temperature and pressure tend to promote detonation. The ability of the unburnt mixture to absorb or get rid of the heat radiated to it by the advancing flame front is also important. The latent enthalpy of the mixture and the design of the combustion chamber affect this ability. The latter must be arranged for adequate cooling of the unburnt mixture by placing it near a well cooled feature such as an inlet valve.

Key fact
At higher speeds detonation can be extremely dangerous, prompting pre-ignition and possibly the complete wreckage of the engine.

The length of flame travel should be kept as short as possible by careful positioning of the point of ignition. Other factors include the time (hence ignition timing), since the reaction in the unburnt mixture take some time to develop; the degree of turbulence (in general higher turbulence tends to reduce detonation effects), and most importantly the tendency of the fuel itself to detonate.

Some fuels behave better in this respect. Fuel can be treated by additives (e.g. tetra-ethyl lead) to improve performance. However, this aggravated an already difficult pollution problem. A fuel with good anti-knock properties is isooctane, and a fuel that is susceptible to detonation is normal heptane.

To obtain the octane number or the anti-knock ratings of a blend of fuel, a test is carried out on an engine run under carefully monitored conditions and the onset of detonation compared with those obtained with various mixtures of isooctane and normal heptane. If the performance of the fuel is identical to, for example, a mixture of 90% isooctane and 10% heptane the fuel is said to have an octane rating of 90.

Definition
If the performance of the fuel is identical to, for example a mixture of 90% isooctane and 10% heptane the fuel is said to have an octane rating of 90.

Mixing water or methanol and water with the fuel can reduce detonation. A mainly alcohol fuel (ethanol for example), which enables the water to be held in solution, is also helpful because then better use can be made of the latent enthalpy of the water.

The evidence of the presence of pre-ignition is not as apparent at the onset as at detonation, but the results are far more serious. There is no characteristic 'ping'; if audible at all, it sounds like a dull thud. Since it is not immediately noticeable its effects are often allowed to take a serious toll on the engine. The process of combustion is not affected to any extent, but a serious factor is that the control of ignition timing can be lost.

Pre-ignition can occur at the time of the spark with no visible effect. More seriously this 'auto-ignition' may creep in earlier in the cycle. The danger of pre-ignition lies not so much in development of high pressures but in the very great increase in heat flow to the piston and cylinder walls. The maximum pressure in fact does not increase appreciably although it may occur a little early.

Pre-ignition is often initiated by some form of hot spot, perhaps red-hot carbon or some poorly cooled feature of combustion space. In some cases, if the incorrect spark plug is used overheated electrodes are responsible, but often detonation is the prime cause. The detonation wave scours the cylinder walls of residual gases present in a film on the surface with the result that the prime source of resistance to heat flow is removed and a great release of heat occurs.

5.3.3 Mixture strength and performance

Figure 5.18 shows the effect of operating at part throttle with varying mixture strength. The chemically correct mixture of approximately 14.7:1 lies between the ratio, which provides maximum power 12:1, and minimum consumption 16:1. The stoichiometric ratio of 14.7:1 is known as a lambda value of one.

A very weak mixture is difficult to ignite but can reduce emissions and improve economy. One technique to get around the problem of igniting weak mixtures is stratification.

It is found that if the mixture strength is increased near the plug and weakened in the main combustion chamber an overall reduction in mixture strength results, but with a corresponding increase in thermal efficiency. Direct injection systems, where fuel is injected into the cylinder instead of the inlet manifold, can be used to achieve stratification.

> **Key fact**
> A very weak mixture is difficult to ignite but can reduce emissions and improve economy.

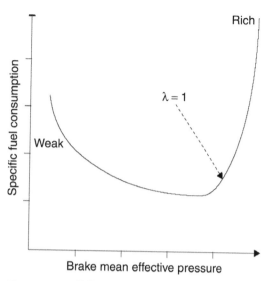

Figure 5.18 Effect of varying the mixture strength while maintaining the throttle position, engine speed and ignition timing

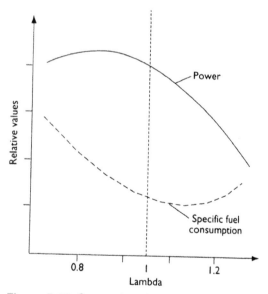

Figure 5.19 Comparison between engine power output and fuel consumption with changes in air-fuel ratio

5.3.4 Compression ignition engines

The process of combustion in the compression ignition engine differs from that in a spark ignition engine. In this case the fuel is injected in a liquid state, into a highly compressed, high temperature air supply in the engine cylinder. Each minute droplet is quickly surrounded by an envelope of its own vapour as it enters the highly heated air. This vapour, after a certain time, becomes inflamed on the surface. A cross-section of any one droplet would reveal a central core of liquid, a thin surrounding film of vapour, with an outer layer of flame. This sequence of vaporisation and burning persists if combustion continues.

The process of combustion (oxidisation of the hydrocarbon fuel), is a lengthy process, but one that may be accelerated artificially by providing the most suitable conditions. The oxidisation of the fuel will proceed in air at normal atmospheric temperatures, but it will be greatly accelerated if the temperature is raised. It will take years at 20°C, a few days at 200°C and just a few minutes at 250°C. In these cases, the rate of temperature rise due to oxidisation is less than the rate at which the heat is being lost due to convection and radiation. Ultimately, as the temperature is raised, a critical stage is reached where heat is being generated by oxidisation at a greater rate than it is being dissipated.

> **Key fact**
> The process of combustion is the oxidisation (burning) of the hydrocarbon fuel.

The temperature then proceeds to rise automatically; this in turn speeds up the oxidisation process and with it the release of heat. Events now take place very rapidly; a flame is established, and ignition takes place. The temperature at which this critical change takes place is usually termed the self-ignition temperature of the fuel. This however depends on many factors such as pressure, time and the ability to transmit heat from the initial oxidisation.

At a temperature well above the ignition point, the extreme outer surface of the droplet of fuel, injected into a hot cylinder, immediately starts to evaporate and this surrounds the core with a thin film of vapour. This involves heat from the air surrounding the droplet to supply the latent enthalpy of evaporation. This is maintained by continuing to draw on the main supply of heat, from the mass of hot air.

> **Key fact**
> At a temperature well above the ignition point, the extreme outer surface of the droplet of fuel, injected into a hot cylinder, immediately starts to evaporate.

Ignition can and will occur on the vapour envelope even with the core of the droplet still liquid and relatively cold. Once the flame is established the combustion proceeds at a more rapid rate. This causes a delay period, after injection commences and before ignition takes place. The delay period therefore depends on:

- ▶ Excess of air temperature over and above the self-ignition temperature of the fuel.
- ▶ Air pressure, both from the point of view of the supply of oxygen and improved heat transfer between the hot air and cold fuel.

Once the delay period is over the rate at which each flaming droplet can find fresh oxygen, to replenish its consumption, controls the rate of further burning. The relative velocity of the droplet to the surrounding air is thus of considerable importance. In the compression ignition engine, the fuel is injected over a period of perhaps 40–50° of crank angle. This means that the oxygen supply is absorbed by the fuel first injected, with a possible

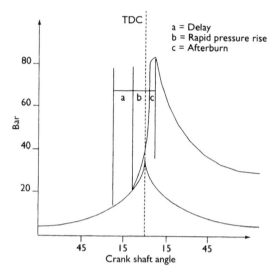

Figure 5.20 Combustion phases in a CI engine

starvation of the last fuel injected. This necessitates a degree of turbulence of the air so that the burnt gases are scavenged from the injector zone and fresh air is brought into contact with the fuel. The turbulence should be orderly and not disorganised as in a spark ignition engine where it is only necessary to break up the flame front.

In a compression ignition engine the combustion can be regarded as occurring in three distinct phases, as shown in Figure 5.20.

1 Delay period
2 Rapid pressure rise
3 After burning, the fuel burning as it leaves the injector.

The longer the delay, the greater and more rapid the pressure rise, since more fuel will be present in the cylinder before the rate of burning comes under direct control of the rate of injection. The aim should be to reduce the delay as much as possible, for the sake of smooth running, the avoidance of knock and to maintain control over the pressure change. There is however a lower limit to the delay since without delay all the droplets would burn as they leave the nozzle. This would

make it almost impossible to provide enough combustion air within the concentrated spray, and the delay period also has its use in providing time for the proper distribution of the fuel. The delay period therefore depends on:

▶ The pressure and temperature of the air
▶ The cetane rating of the fuel
▶ The volatility and latent enthalpy of the fuel
▶ The droplet size
▶ Controlled turbulence.

Key fact

In diesel combustion the aim is to reduce the delay period as much as possible.

The effect of droplet size is important as the rate of droplet burning depends primarily on the rate at which oxygen becomes available. It is however vital for the droplet to penetrate some distance from the nozzle around which burning will later become concentrated. To do this the size of the droplets must be large enough to obtain sufficient momentum at injection. On the other hand, the smaller the droplet the greater the relative surface area exposed and the shorter delay period. A compromise between these two effects is clearly necessary.

With high compression ratios (15:1 and above) the temperature and pressure are raised so that the delay is reduced. High compression ratios are however a disadvantage mechanically and inhibit the design of the combustion chamber, particularly in small engines where the bumping clearance consumes a large proportion of the clearance volume.

Key fact

With high compression ratio engines (15:1), the temperature and pressure are increased.

5.3.5 Summary of combustion

The previous sections looked at some of the issues of combustion. This is intended to provide background to help with the understanding of how alternative fuels are handled by modern control systems – and why there is much more to an AF vehicle than switching to a different source of fuel! Some of the key issues this section has raised are the time to burn a fuel air mixture, the effects of changes in mixture strength and ignition timing, the consequences of detonation and other design challenges.

Different types of fuel burn in different ways and hence it is essential to make sure that mixture strength, timing (ignition and injection), temperatures and more, are set to an optimum value.

> **Key fact**
> Accurate control of engine operating variables is the key way to control the combustion process for all types of fuel.

5.4 Ignition systems

5.4.1 Introduction

For a spark to jump across an air gap of 1.0 mm under normal atmospheric conditions (1 bar), a voltage of 4 to 5 kV is required. For a spark to jump across a similar gap in an engine cylinder, having a compression ratio of 8:1, approximately 10 kV is required. For higher compression ratios and weaker mixtures, a voltage up to 20 kV may be necessary. The ignition system transforms the normal battery voltage of 12 V to approximately 8 to 20 kV and, in addition, must deliver this high voltage to the right cylinder, at the right time. Some ignition systems supply up to 40 kV to the spark plugs.

> **Key fact**
> The purpose of the ignition system is to supply a spark inside the cylinder, near the end of the compression stroke, to ignite the compressed charge of air/fuel vapour.

The fundamental operation of most ignition systems is similar. One winding of a coil is switched on and off causing a high voltage to be induced in a second winding.

5.4.2 Systems

Modern ignition systems now are part of the engine management, which controls fuel delivery, ignition, and other vehicle functions. The main ignition components are the engine speed and load sensors, knock sensor, temperature sensor and the ignition coil. The ECU reads from the sensors, interprets and compares the data, and sends output signals to the actuators. The output component for ignition is the coil and then the spark plugs.

> **Key fact**
> Modern ignition systems now are part of the engine management.

Ignition systems continue to develop and will continue to improve. However, keep in mind that the simple purpose of an ignition system is to ignite the fuel air mixture every time at the right time. And, no matter how complex the electronics may seem, the high voltage is produced by switching a coil on and off.

> **Key fact**
> No matter how complex the electronics may seem, a high voltage spark is produced by switching a coil on and off.

If two coils (known as the primary and secondary) are wound on to the same iron core, then any change in magnetism of one coil will induce a voltage into the other. This happens when a current is switched on and off to the primary coil. If the number of turns of wire on the secondary coil is more than the primary a higher voltage can be produced. This is called transformer action and is the principle of the ignition coil.

Safety first

Ignition systems operate at very high voltage.

Direct ignition, or coil on plug (COP) ignition, is in common use and utilises an inductive coil for each engine cylinder. These coils are mounted directly on the spark plugs. The use of an individual coil for each plug ensures that the charge time is very fast (full coil charge in a very small dwell angle). This ensures that a very high voltage, high-energy spark is produced. This voltage, which can be more than 40 kV, provides efficient initiation of the combustion process under cold starting conditions and with weak mixtures.

Key fact

Direct ignition has a coil for each spark plug.

5.4.3 Timing

For optimum efficiency the ignition advance angle should be such as to cause the maximum combustion pressure to occur at about 10° after TDC. The ideal ignition timing is dependent on two main factors, engine speed and engine load. An increase in engine speed requires the ignition timing to be advanced. The cylinder charge, of air fuel mixture, requires a certain time to burn (about

Figure 5.21 COP ignition coil. (Source: Denso)

2 ms). At higher engine speeds the time taken for the piston to travel the same distance reduces. Advancing the time of the spark ensures full burning is achieved.

Key fact

For optimum efficiency the ignition advance angle should be such as to cause the maximum combustion pressure to occur about 10° after TDC (but this will vary between types of engine and types of fuel).

A change in timing due to engine load is also required as the weaker mixture used on low load conditions burns at a slower rate. In this situation further ignition advance is necessary. Greater load on the engine requires a richer mixture, which burns more rapidly. In this case some retardation of timing is necessary. Absolute manifold pressure (MAP) is proportional to engine load. Overall, under any condition of engine speed and load an ideal advance angle is required to ensure maximum pressure is achieved in the cylinder just after TDC. The ideal advance angle may also be determined by engine temperature and any risk of detonation.

All current systems are known as constant energy ensuring high performance ignition

Figure 5.22 Manifold pressure sensor

Figure 5.23 Inductive sensor

even at high engine speed because the dwell is varied.

A crankshaft position sensor (CPS) is used to determine engine speed and sometimes position (Figure 5.22). It is usually an inductive sensor and is positioned against the front of the flywheel or against a reluctor wheel just behind the front crankshaft pulley. Hall effect sensors are also used.

On most cars the ignition system is combined with the fuel system so that even more accurate control of outputs is possible and input data from sensors can be shared. Ignition timing and dwell are controlled by an electronic control unit (ECU) that uses information from sensors. Typically, these are: engine speed, engine load, engine temperature and sometimes engine knock sensor.

Inside the ECU data is stored in the form of look up tables, so that at a certain speed and certain load, for example, a set timing figure is used. These tables can be represented by a cartographic map, as shown in Figure 5.24.

Many more 'corrections' are made to ignition timing after the base value has been 'looked up'. An example screen grab from a computer program that allows the manipulation of this data on some cars is shown as Figure 5.33. There are several look up tables shown, just for ignition timing.

5.4.4 Spark plugs

The simple requirement of a spark plug is that it must allow a spark to form within the combustion chamber, to initiate combustion. To do this the plug has to withstand severe conditions in the cylinder. It must withstand severe vibration and a harsh chemical environment. Finally, but perhaps most importantly, the insulation properties must withstand voltages pressures up to 40 kV.

Figure 5.25 shows a range of spark plugs. The centre electrode is connected to the top terminal by a stud. The electrode is

Base ignition timing map

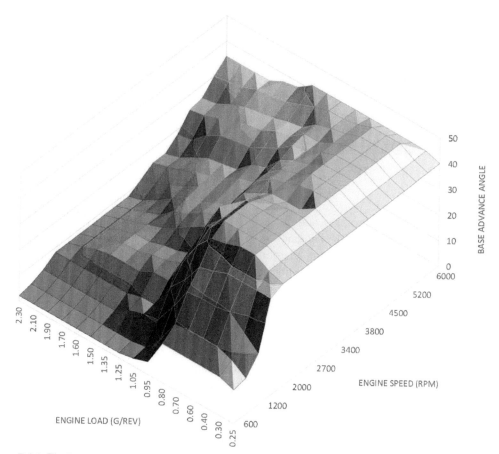

Figure 5.24 Timing map

constructed of a nickel-based alloy, silver or platinum. If a copper core is used in the electrode this improves the thermal conduction properties. The insulating material is ceramic based and of a very high grade. Flash over or tracking down the outside of the plug insulation is prevented by ribs that effectively increase the surface distance from the terminal to the metal fixing bolt, which is of course earthed to the engine.

Key fact

The ideal operating temperature of a spark plug electrode is between 400 and 900°C.

The heat range of a spark plug is a measure of its ability to transfer heat away from the centre electrode. A hot running engine will require plugs with a higher thermal ability than a colder running engine. Note that hot and cold running of an engine in this sense refers to the combustion temperature, not to the cooling system.

Spark plug electrode gaps in general have increased as the power of the ignition systems driving the spark has increased. The simple relationship between plug gap and voltage required is that as the gap increases so must the voltage (leaving aside engine

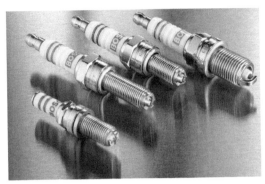

Figure 5.25 A range of spark plugs. (Source: Bosch Media)

operating conditions). Further, the energy available to form a spark at a fixed engine speed is constant, which means that a larger gap using higher voltage will result in a shorter duration spark. A smaller gap will allow a longer duration spark.

For cold starting an engine and for igniting weak mixtures the duration of the spark is critical. Likewise, the plug gap must be as large as possible to allow easy access for the mixture to prevent quenching of the flame. The final choice is therefore a compromise reached through testing and development.

> **Key fact**
> Plug gaps of 0.6 to 1.2 mm are in common use.

5.5 Fuel systems

5.5.1 Introduction

The ideal air–fuel ratio for burning petrol/gasoline, is about 14.7:1. This is the theoretical amount of air required to completely burn the fuel. It is given a lambda (λ) value of 1.

> **Definition**
> **Lambda (λ) value of 1**: The ideal theoretical air–fuel ratio for burning petrol/gasoline, which is about 14.7:1.

Air–fuel ratio is altered during the following operating conditions of an engine to improve its performance, driveability, consumption and emissions:

▶ Cold starting – a richer mixture is needed to compensate for fuel condensation and improves driveability.
▶ Load or acceleration – richer to improve performance.
▶ Cruise or light loads – weaker for economy.
▶ Overrun – very weak (if any fuel) to improve emissions and economy.

The more accurately the air–fuel ratio is controlled to cater for external conditions, then the better the overall operation of the engine. The ratio is controlled by changing the amount of fuel injected during the four-stroke cycle.

Diesel fuel systems are controlled in a similar manner other than they involve much higher pressures and different designs for the main components.

All systems are now electronically controlled, whether they are petrol/gasoline or diesel. This allows the operation of the fuel system to be very closely matched to the requirements of the engine and type of fuel. This matching process is carried out during development on test beds and dynamometers, as well as development in the car. The ideal operating data for all engine operating conditions is stored in a read only memory in the ECU just like the ignition timing mentioned earlier. Close control of fuel quantity injected allows the optimum setting for mixture strength under all operating conditions.

Key fact

The main advantage of a petrol/gasoline fuel injection system is accurate control of the fuel quantity injected into the engine.

Key fact

All modern engine management systems are electronically controlled.

5.5.2 Petrol injection systems

The main advantage of a petrol/gasoline fuel injection system is accurate control of the fuel quantity injected into the engine. The basic principle of fuel injection is that if petrol is supplied to an injector (electrically controlled valve), at a constant differential pressure, then the amount of fuel injected will be directly proportional to the injector open time.

The two most important inputs to the electronic control system are speed and load. The basic fuelling requirement is determined from these inputs in a similar way to ignition timing. An engine's fuelling requirements are stored as part of a read only memory (ROM) chip in the ECU. When the ECU has determined the 'look up value' of the fuel required (injector open time), corrections to this figure can be added for battery voltage,

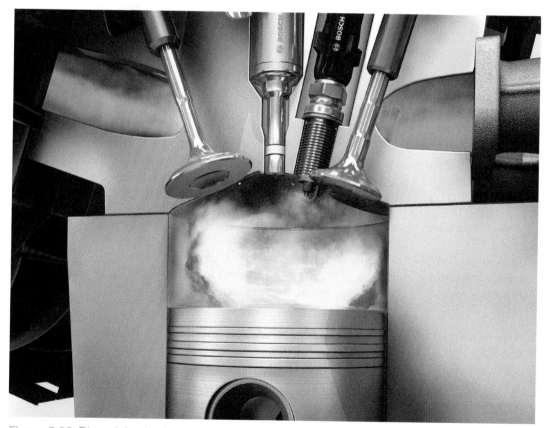

Figure 5.26 Direct injection injector in a cylinder. (Source: Bosch Media)

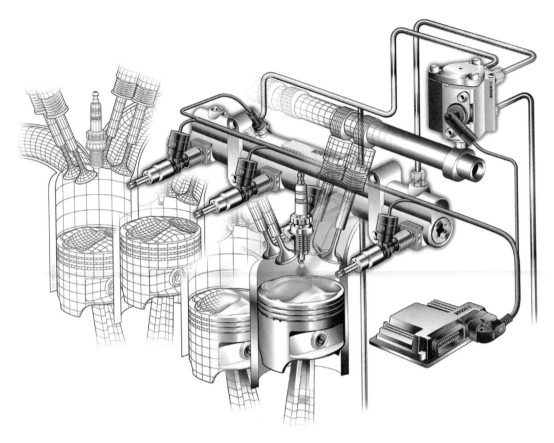

Figure 5.27 Direct injection system and components. (Source: Bosch Media)

temperature, throttle change or position and fuel cut-off. Figure 5.27 shows an injection system and ECU.

Key fact

The two most important inputs to the electronic control unit (ECU) and engine speed and load.

Idle speed and fast idle are also generally controlled by the ECU and a suitable actuator. It is also possible to have a form of closed loop control with electronic fuel injection. This involves a lambda sensor to monitor exhaust gas oxygen content. This allows very accurate control of the mixture strength, as the oxygen

content of the exhaust is proportional to the air fuel ratio. The signal from the lambda sensor is used to adjust the injector open time.

Definition

Closed loop control: A system where the outputs have a direct effect on the inputs.

A fuel pump is used to supply pressurised fuel to the injectors. The type varies depending on whether the fuel is injected into the inlet manifold or directly into the engine cylinders. In both cases, the pump ensures a constant supply of fuel to the fuel rail. A pressure regulator keeps the pressure constant so that

Figure 5.28 Lambda sensor

Figure 5.29 Injector

the quantity injected is only determined by injection duration. The volume in the rail acts as a swamp to prevent pressure fluctuations as the injectors operate.

The pressurised fuel supplied to the injectors must be free from any contamination or else the injector nozzles will be damaged or blocked. A filter is used to prevent this.

The temperature sensor used to determine the engine coolant temperature is a thermistor.

Definition

Thermistor: Electrical resistor whose resistance is reduced as temperature is increased.

Fuel injectors are simple solenoid operated valves designed to operate very quickly and produce a finely atomised spray pattern.

A throttle position sensor is used to supply information as to whether the throttle is at idle, full load or somewhere in between. Many cars do not have a throttle cable but instead use an electrical actuator controlled by the ECU.

The quantity of fuel to be injected is determined primarily by the quantity of air

drawn into the engine. This is dependent on two factors:

▶ Engine speed (rpm)
▶ Engine load (inlet manifold pressure).

The load sensor (Figure 5.22) is connected to the manifold by a pipe, and senses manifold absolute pressure. The other method of sensing engine load is direct measurement of air intake quantity using a hot wire meter or a flap type air flow meter (Figure 5.31).

To operate the injectors, the ECU needs to know, in addition to air pressure or air flow, the engine speed to determine the injection quantity. The same flywheel sensor used by the ignition system provides this information. The injectors either operate simultaneously once per engine revolution, injecting half of the required fuel, or they are operated sequentially in the firing order.

Key fact

Modern direct injections systems operate the injectors sequentially.

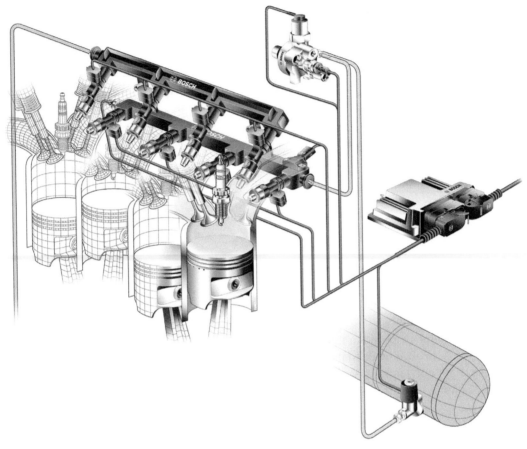

Figure 5.30 Dual fuel (petrol and CNG) injection system. (Source: Bosch Media)

Figure 5.31 Hot wire air flow meter. (Source: Bosch Media)

A basic open period for the injectors is determined by using the ROM information relating to engine load and engine speed (an example represented as a cartographic map is shown as Figure 5.32). Corrections are then made relative to air temperature and whether the engine is idling, or at full or partial load. The ECU then carries out another group of corrections, if applicable:

▶ After start enrichment
▶ Operational enrichment
▶ Acceleration enrichment
▶ Weakening on deceleration
▶ Cut-off on over run
▶ Reinstatement of injection after cut-off
▶ Correction for battery voltage variation.

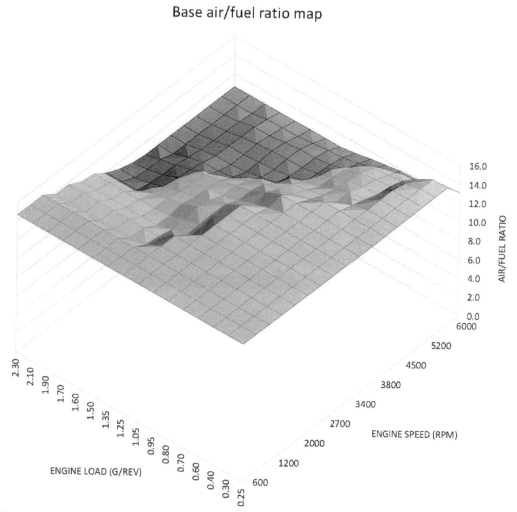

Figure 5.32 Base air/fuel ratio map (lambda value)

There are many other corrections that can be made, and these depend on the complexity of the system. An example screen grab from a computer program that allows the manipulation of this data on some cars is shown as Figure 5.33.

5.5.3 Diesel injection systems

Diesel engines have the fuel injected into the combustion chamber where it is ignited by heat in the air charge. This is known as compression ignition (CI) because no spark is required. The high temperature needed to ignite the fuel is obtained by a high compression of the air charge. Diesel fuel is injected under high pressure from an injector nozzle, into the combustion chambers. The fuel is pressurised in a diesel injection pump. It is supplied and distributed to the injectors through high pressure fuel pipes or directly from a rail and/or an injector. The high-pressure generation is sometimes created by a direct acting cam on each injector, but mostly on smaller engines, by a separate

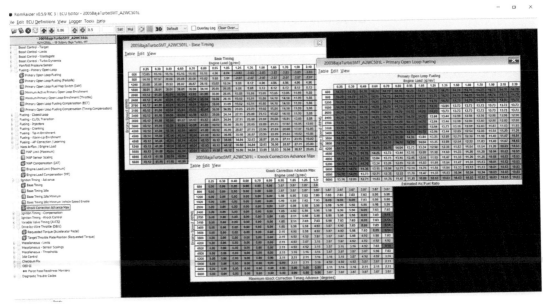

Figure 5.33 Look up table data examples

pump. A range of diesel injection components is shown as Figure 5.34.

> **Key fact**
>
> The high temperature needed to ignite fuel in a diesel engine is obtained by high compression of the air charge.

The air flow into a diesel engine is usually unobstructed by a throttle plate so a large air charge is always provided. Throttle plates may be used to provide control for emission devices. Engine speed is controlled by the amount of fuel injected. The engine is stopped by cutting off the fuel delivery. For all engine operating conditions, a surplus amount of air is needed for complete combustion of the fuel.

Earlier diesel engines tended to be considered as indirect or direct injection. Almost all are direct, and these are shown as Figure 5.35.

Small high-speed diesel engine compression ratios are about 20:1 for direct injection

systems. This compression ratio can raise the air charge to temperatures of between 500°C and 800°C. Very rapid combustion of the fuel therefore occurs when it is injected into the hot air charge.

To aid starting and to reduce diesel knock, cold start devices may be used. For early indirect injection engines, starting at lower than normal operating temperatures requires additional combustion chamber heating. For direct injection engines, cold start devices are only required in frosty weather.

Ignition of the fuel occurs in the combustion chamber at the time of injection into the heated air charge. The injection point and the ignition timing are therefore effectively the same thing. Diesel engine injection timing is equivalent to the ignition timing for petrol engines. Injection timing must fall within a narrow angle of crankshaft rotation. It is advanced and retarded for engine speed and load conditions. Incorrect timing leads to power loss. An increase in the production of nitrogen oxides (NOx) when too far advanced

Figure 5.34 Diesel fuel injection components. (Source: Bosch Media)

Rotary Pump   Common-Rail   Unit Injector   Unit Pump

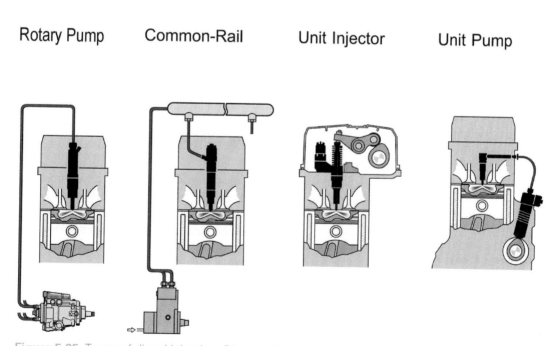

Figure 5.35 Types of diesel injection: Direct using a rotary pump, common rail, unit injection and pumped unit injection. (Source: Bosch Media)

or an increase in hydrocarbon (HC) emissions, when too far retarded also occurs.

The main components of a diesel fuel system provide for either the low-pressure or the high-pressure functions. The low-pressure components are the fuel tank, the fuel feed and return pipes and hoses, a renewable fuel filter with a water trap and drain tap, and a priming or lift pump. The high-pressure components are the fuel injector pump, the high-pressure pipes and the injectors. Other components provide for cold engine starting. Electronically controlled systems include sensors, an electronic diesel control (EDC) module and actuators in the injection pump.

All diesel fuel entering the injection pump and injectors must be fully filtered. The internal components of the pump and injectors are manufactured to very fine tolerances. Even very small particles of dirt could be damaging to these components.

The high-pressure pipes are of double thickness steel construction and are all the same length. This is so that the internal pressure rise characteristics are identical for all cylinders. The high-pressure connections are made by rolled flanges on the pipe ends and threaded unions securing the rolled flanges to convex, or occasionally concave, seats in the delivery valves and injectors.

The development of diesel fuel systems is continuing, with many new electronic changes to the control and injection processes. Common rail (CR) systems, which operate at very high injection pressures, are now used on most small engines.

The common rail system developments have resulted in significant improvements in fuel consumption and performance.

Fuel injection pressures are varied, throughout the engine speed and load range, to suit the instantaneous conditions of driver demand and engine speed and load conditions. The electronic diesel control (EDC) module carries out calculations to determine the quantity of fuel delivered. It also determines the injection timing based on engine speed and load conditions. The actuation of the injectors, at a specific crankshaft angle (injection advance), and for a specific duration (fuel quantity), is controlled by the module.

The high-pressure CR pump is driven from the engine crankshaft through a geared drive at half engine speed. It can also be fitted on the end of the camshaft housing and be driven by the camshaft. It is lubricated by the diesel fuel that flows through it.

The pump is a triple piston radial pump, with a central cam for operation of the pressure direction of the pistons and return springs to maintain the piston rubbing shoes in contact with the cam. The pump has a positive displacement with inlet and outlet valves

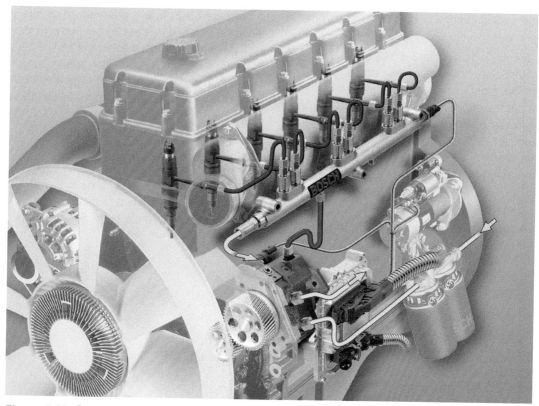

Figure 5.36 Common rail system. (Source: Bosch Media)

① Air mass meter

② Engine ECU

③ High pressure pump

④ Common rail

⑤ Injectors

⑥ Engine speed
sensor

⑦ Coolant temp.
sensor

⑧ Filter

⑨ Accelerator
pedal sensor

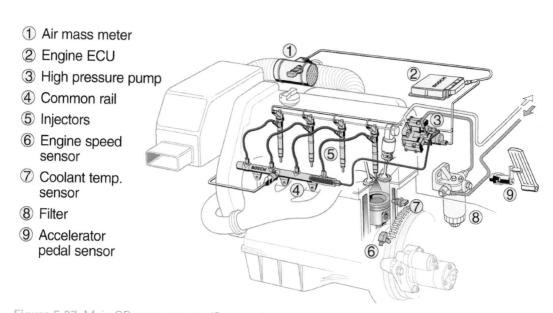

Figure 5.37 Main CR components. (Source: Bosch Media)

controlling the direction of flow through the pump.

The pump delivery rate is proportional to the speed of rotation of the engine so that it meets most engine speed requirements. To meet the engine load requirements, the pump has a high volume. To meet the high-pressure requirements, for fine atomisation of the fuel on injection, the pump can produce pressures in the region of 1400 bar.

Safety first

A CR pump can produce pressures in the region of 1400 bar.

A pressure control valve is a mechanical and electrical unit and is fitted on the pump or the high-pressure rail. The mechanical part of the valve consists of a compression spring that acts on a plunger and ball valve. The electrical component is a solenoid that puts additional and variable force to the ball valve. The solenoid is actuated on signals from the EDC module.

The rail is an accumulator because it holds a large volume of fuel under pressure. The volume of fuel is sufficient to dampen the pressure pulses from the high-pressure pump.

Figure 5.38 Diesel ECU. (Source: Bosch Media)

Definition

Accumulator: an apparatus by means of which energy can be stored (the rail on a CR system).

Key fact

The high-pressure rail is common to all cylinders and is where the system derives its name: common rail (CR).

Opening and closing of the injector is controlled, not by high pressure fuel pulses from an injector pump as in a conventional rotary distributor pump, but by actuation of an electrical solenoid in the injector body. This is controlled by the electronic diesel control module. A permanent high pressure is maintained in the injector at the same pressure as the rail. Operation of the injector is controllable for very small intervals of time.

The electronic control of the common rail diesel injection system allows for precise control of fuelling. This results in excellent economy and very low emissions. The ideal outputs are 'mapped' in the same way as described for petrol systems earlier, and a range of sensors provides inputs to the ECU.

CHAPTER 6

Case studies

6.1 Introduction

This chapter will present some of the technologies used to make alternative fuels viable. Most manufacturers' systems are similar but as always you should refer to their technical information before carrying out any work on their vehicles.[1]

> **Safety first**
>
> Always refer to specific manufacturers' technical information before carrying out any work on their vehicles.

6.2 Volkswagen Golf BiFuel LPG

6.2.1 Overview

On this vehicle, like most, all components that are required for gas operation are factory fitted. Normal petrol/gasoline operation is available as with any other version. The additional components needed for the liquid petroleum gas (LPG) system are the:

- gas filler neck
- LPG tank

- selection button with the gas supply gauge and the fuel selection switch
- vaporiser
- gas filter
- gas fuel rail
- gas injection valves
- gas rail sensor
- gas ECU.

Because the engine is based on the FlexFuel (E85) engine (already modified as discussed earlier), which is based on the technology from the 1.6 L 75 kW BSE engine, no additional mechanical modifications are required for LPG operation. Normal petrol operation is still possible, and works like on any other similar vehicle. LPG is simply selected by a button in the centre console. The engine can achieve 72 kW in gas mode.

6.2.2 Gas filler

The gas filler neck is next to the petrol filler neck behind the tank filler flap. It is connected to the LPG tank via a pipe. The gas filler neck has a check valve (Figure 6.1), which allows the liquid gas to only flow in one direction (in to the tank). The valve opens as the tank is filled under pressure.

Alternative Fuel Vehicles. 978-1-138-50370-0 © 2018 Tom Denton.
Published by Taylor & Francis. All rights reserved.

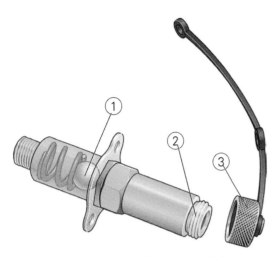

Figure 6.1 Gas filler: 1. Pressure relief check valve, 2. Filler neck, 3. Cap. (Source: Volkswagen)

and has crash-optimised mounts for safety. It includes a swirl pot, a pressure relief valve, an automatic fill limiter, a gas gauge sender and a gas tank valve. The automatic fill limiter stops the refuelling process when there is 80% LPG in the tank. This is because the filling level can fluctuate due to different temperatures. At a temperature of 15°C, for example, the tank holds 39 litres. The automatic tank fill limiter uses a simple float and cam mechanism to close a valve, which prevents overfilling.

> **Key fact**
> LPG filling level can fluctuate due to different temperatures.

The spring loaded, and solenoid operated, gas tank valve (Figure 6.2) is fitted in the valve pot and is used to switch the gas supply on and off as commanded by the ECU. The valve closes automatically so no liquid gas flows to the vaporiser when in petrol mode. It also closes when the engine is turned off, in an accident, or if the supply voltage fails.

A pressure relief valve (Figure 6.3) is fitted in the LPG tank valve pot, which is to prevent the tank bursting if the pressure rises excessively. This could happen in very high temperatures. If pressure in the tank reaches 27.5 bar, the valve opens mechanically. The LPG is then vented away from the passenger compartment via plastic breather hoses. A plastic dust cap is push off as the valve opens.

6.2.4 Vaporiser

The liquid petroleum is converted to a gaseous state in a vaporiser (Figure 6.4), which also reduces the pressure from about 10 bar to 1 bar above the pressure in the intake manifold. As with petrol injection systems, this is sometimes described as a differential pressure. LPG is expanded in the vaporiser in two stages, which allows better compensation of pressure fluctuations.

Three different connection systems are currently in use across Europe for filling LPG vehicles. These are the:

▶ ACME connector
▶ Dish connector
▶ Bayonet connector.

The country the vehicle is being used in determines which connector is needed to connect to LPG pumps.

> **Key fact**
> Different LPG connection systems are currently in use across Europe for filling vehicles.

6.2.3 Pipes and valves

The LPG system is divided into a high-pressure and a low-pressure area. The LPG system uses copper with PVC sleeving for the high-pressure pipes and special plastic hoses for the low-pressure pipes.

The 49-litre tank is made from 3.5 mm thick steel and is fitted in the spare wheel well,

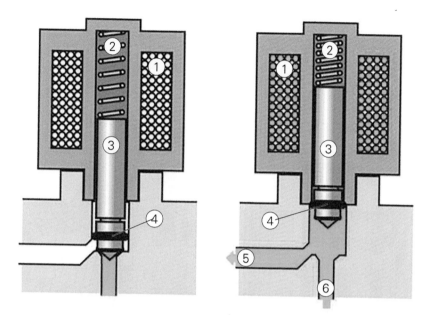

Figure 6.2 Tank valve: 1. Coil, 2. Spring, 3. Plunger, 4. Valve, 5. To vaporiser, 6. From tank. (Source: Volkswagen)

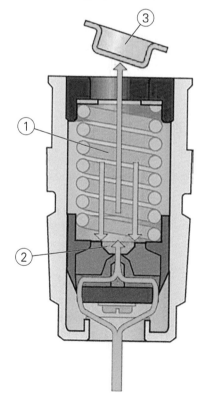

Figure 6.3 Pressure relief valve: 1. Spring, 2. Valve, 3. Cap. (Source: Volkswagen)

> **Key fact**
>
> The vaporiser is a key component of an LPG fuel system.

Referring to Figure 6.4, each stage of the vaporiser consists of:

- an inner chamber (5 and 21)
- an outer chamber (12 and 16)
- a control chamber (10 and 17).

LPG passes from the first stage to the second stage of the vaporiser, through the overflow channel (2). Each stage has a valve with a flap and a plunger. The plunger is bolted to the diaphragm (4 and 23) and there is a spring in each spring chamber (9 and 18). Atmospheric pressure is present in the spring chamber of the first stage. Inlet manifold pressure is present in the spring chamber of the second stage. A rubber seal (14) between the first and second stage separates the cooling circuit.

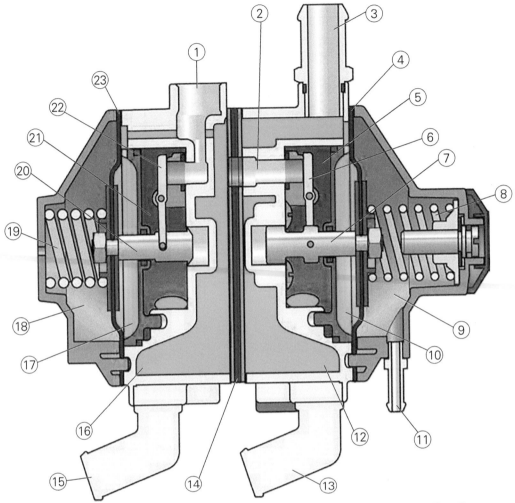

Figure 6.4 Vaporiser: 1. Supply line from high pressure valve for gas mode, 2. Overflow channel, 3. Outlet to gas filter, 4. Diaphragm, 5. Inner chamber, 6. Flap, 7. Plunger, 8. Spring, 9. Spring chamber, 10. Control chamber, 11. Vacuum connection, 12. Outer chamber, 13. Coolant outlet, 14. Rubber seal, 15. Coolant inlet, 16. Outer chamber, 17. Control chamber, 18. Spring chamber, 19. Spring, 20. Plunger, 21. Inner chamber, 22. Flap, 23. Diaphragm. (Source: Volkswagen)

6.2.5 Gas fuel rail and injectors

The gas fuel rail, with an integrated pressure and temperature sensor, is mounted on the engine inlet manifold. Four electrically controlled gas injection valves are also fitted; one for each cylinder (Figure 6.5).

The LPG from the gas filter flows into the gas fuel rail. The ECU switched the gas injectors on for a set duration and, because the pressure is controlled, a metered quantity of gas passes into the inlet manifold through a plastic pipe.

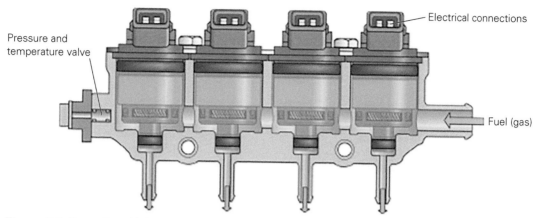

Pressure and temperature valve

Electrical connections

Fuel (gas)

Figure 6.5 Gas rail and injectors with the pressure and temperature valve on the left side. (Source: Volkswagen)

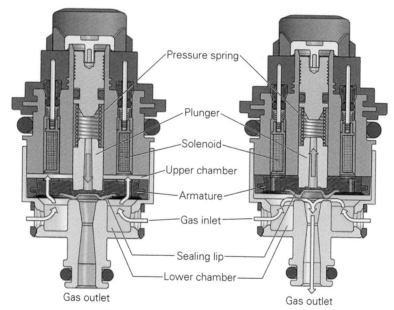

Pressure spring

Plunger

Solenoid

Upper chamber

Armature

Gas inlet

Sealing lip

Lower chamber

Gas outlet

Gas outlet

Figure 6.6 Gas injector in the off (left) and on (right) positions. (Source: Volkswagen)

Key fact
Gas injectors are switched on for a set duration to accurately control the quantity of gas to the engine.

The pressure and temperature sensors supply information for the gas control ECU and signal when it is necessary to switch back to petrol/ gasoline mode. This could be due to:

▶ an empty LPG tank
▶ a gas system pressure drop
▶ a clogged gas filter.

When the injectors (Figure 6.6) are operated by the ECU, the solenoid generates a magnetic field and the armature, with its sealing lip, is pulled against the force of the pressure spring. The LPG from the upper chamber then flows via the holes in the armature into the lower chamber, and

75

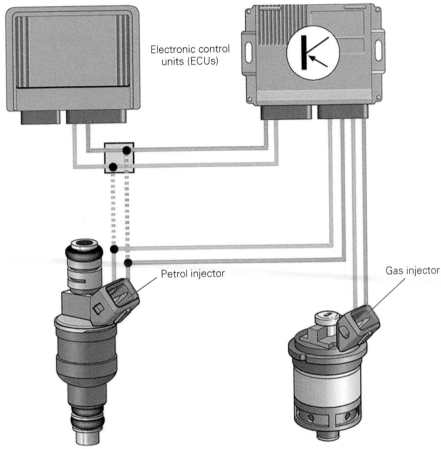

Electronic control
units (ECUs)

Petrol injector

Gas injector

Figure 6.7 Petrol and gas ECUs share some connections. (Source: Volkswagen)

through the pipes and intake manifold into the combustion chamber.

6.2.6 Control units

In addition to the engine control unit, a second unit is required to control the gas mode. There is a connector on the engine wiring harness running to the petrol injectors as shown in Figure 6.7. The petrol injection signals are interrupted there and forwarded to the gas mode control unit. This signal is used to calculate the gas injection times and, to avoid a fault code entry in the engine control unit, it receives the expected petrol injector signals via resistors in the gas mode control unit.

Figure 6.8 is the general 'input-control-output' diagram for the LPG system. As with all complex control systems, the sensors on the left provide information to the control unit, which, using data stored in memory, operates the actuators on the right.

6.3 Volkswagen fuel cell vehicle

6.3.1 Overview

The Volkswagen HyMotion, like all fuel cell vehicles, is powered by electricity in the same way as a battery-electric car. The drivetrain can therefore rely on the tried and tested

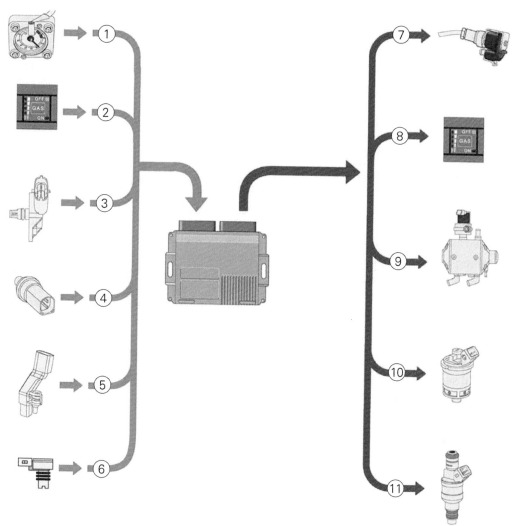

Figure 6.8 Gas ECU with its input sensors: 1. Gas gauge sender, 2. Switch, 3. Gas rail sensor, 4. Coolant temperature sensor, 5. Engine speed sensor, 6. Manifold pressure (load) sensor. Output actuators: 7. Gas tank valve, 8. Gas gauge, 9. High-pressure valve on the vaporiser, 10. Gas injectors, 11. Petrol/Gasoline injectors. (Source: Volkswagen)

e-modules and technologies used by the EVs in the VW range, the eGolf for example.

This short case study highlights some of the key components and features of this electric vehicle, the range of which is about 500 km.

6.3.2 Operation

The system includes a 1.1 kWh lithium-ion battery for storing regenerated energy. A DC-DC converter is then used for setting the voltage, and controlling the flow of energy between the battery, fuel cell and electric motor.

> **Definition**
> **Kilowatt/hour (kWh):** A measure of electrical energy equivalent to a power consumption of 1,000 watts for 1 hour.

The fuel cell stack consists of many individual cells that generate electric power for the electric motor through a chemical reaction of hydrogen and oxygen (see Chapter 4). Other components are needed to operate the stack, for example the turbo compressor and an inlet and outlet control. This is to make sure that the correct amount of air (oxygen), hydrogen and cooling water is fed into the fuel cells.

A radiator is used to cool the stack by releasing the heat produced during the energy conversion into the environment. This fuel cell stack converts about 60% of the energy in the hydrogen into electricity.

The HyMotion uses the same motor as the e-Golf. It can produce up to 270 Nm or torque and 12,000 rpm. Like most pure EVs, the Golf HyMotion uses a single speed transmission. The electric motor drives the front wheels via the transmission and the drive shafts. Top speed is about 160 km/h and the 0–100 km/h time is 10 seconds.

The lithium-ion battery stores the energy recovered during regenerative braking. It also supplies the motor in dynamic phases. The prototype uses a high-performance battery, which can also be charged via the fuel cell.

> **Key fact**
>
> The HyMotion lithium-ion battery stores the energy recovered during regenerative braking.

The power electronics are connected to the electric motor and the battery, and transform the electric current so that the electric motor works as a motor or generator. During normal driving, the power electronics transform the direct current of the high-voltage battery into a three-phase alternating current (using an inverter) that drives the electric motor. In generator mode, it transforms the alternating current into direct current to charge the battery.

Figure 6.9 Volkswagen fuel cell car. (Source: Volkswagen Media)

Four hydrogen fuel tanks are used, and each has an inner plastic and outer carbon fibre layer. They are located underneath the vehicle.

6.4 Ford Transit CNG

6.4.1 Overview

The CNG arrangement on the Ford Transit used for this case study example is a spark-ignition, electronically controlled, sequential injection system. One or two 80-litre tanks are used and filled from special supply stations

to a pressure of 160 to 200 bar. The tanks are designed to withstand pressures of up to 350 bar. As fuel is used, the pressure drops. A manual shut-off valve is incorporated as part of the tank and this must be turned off during repairs. A solenoid controlled valve is switched by the electronic control unit.

> **Key fact**
> The CNG is kept at a pressure between 160 and 200 bar.

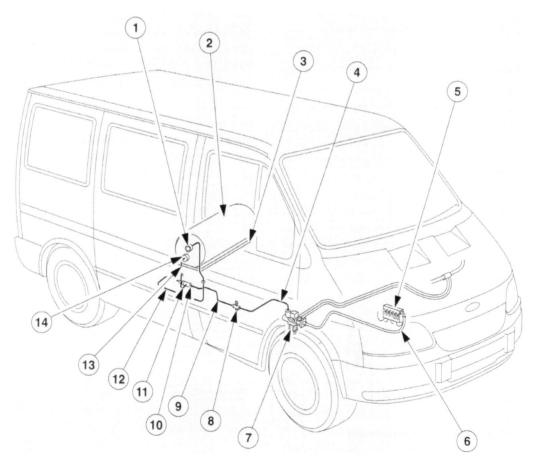

Figure 6.10 Main components: 1. Fuel filler, 2. Fuel tank, 3. Fuel tank mounting frame, 4. High-pressure fuel pipe, 5. Fuel injection supply manifold, 6. Low-pressure fuel hose, 7. Fuel pressure regulator, 8. High-pressure sensor and drain plug, 9. High-pressure fuel pipe, 10. One-way valve, 11. T-piece, 12. High-pressure fuel pipe, 13. High-pressure fuel pipe, 14. Fuel tank valve. (Source: Ford Motor Company)

6.4.2 **Operation**

A fuel regulator is used to maintain gas pressure in the supply manifold. The gas is injected into the inlet manifold using one injector for each cylinder (four in this case). The fuel mixes with air and is combusted in the engine as normal. The regulator incorporates a pressure relief device that operates at 18 bar to protect the low-pressure side. As the CNG is extracted it expands and this has a cooling effect on the pressure regulator (just like the expanding gas does in an AC evaporator). For this reason, the regulator forms part of the engine coolant circuit to ensure that it does not freeze.

> **Key fact**
>
> As CNG is extracted it expands and this causes a cooling effect (like AC operation).

The 'alternative fuel control module' controls the CNG injection in response to information from sensors, just like a normal petrol injection system would do. The alternative fuel control module can communicate with the standard powertrain control module (PCM) so reducing the number of extra sensors needed. Ignition is controlled by the standard PCM, but a different ignition map is used when under CNG operation. Bi-fuel operation allows switching between the two sources and if one runs low than it will switch over automatically.

The engine always starts on petrol but will switch to CNG after a few seconds, if this has been selected by the driver. In extreme cold conditions (−30°C) the switch over may take a few minutes. This ensures easy starting and it also makes sure the petrol injection is used regularly.

A sensor, in the high-pressure CNG line, sends a signal to a digital fuel gauge. If pressure drops (for example when the CNG is running low) the system uses this signal

to switch over to petrol/gasoline operation. After the pressure sensor, the gas flows to the pressure regulator. This reduces the pressure to a constant value of about 9 bar, with respect to intake manifold pressure. The gas now passes to the injectors through a flexible pipe, to allow for engine movement, to the CNG fuel injection supply manifold. This manifold, much like standard petrol injection, incorporates the four injectors. The operation of this system is the same on most similar vehicles, the only real difference being component location.

> **Key fact**
>
> CNG pressure is regulated to a constant value of about 9 bar, with respect to intake manifold pressure.

The fuel filler incorporates a microswitch, which is used to prevent the engine being started while the filler cap is removed. The high-pressure fuel pipes are stainless steel and incorporate a one-way valve to prevent CNG flowing back to the fuel filler.

6.4.3 **Electronic control**

When running on CNG the petrol injectors are switched off, and the CNG injectors are controlled by the alternative fuel control unit. This ECU calculates the injection timing and duration based on information from sensors (Figure 6.12):

> **Key fact**
>
> When this system is running on CNG the petrol injectors are switched off.

▶ Oxygen sensor
▶ Throttle position sensor
▶ Camshaft position sensor
▶ Crankshaft speed sensor
▶ Engine coolant temperature sensor

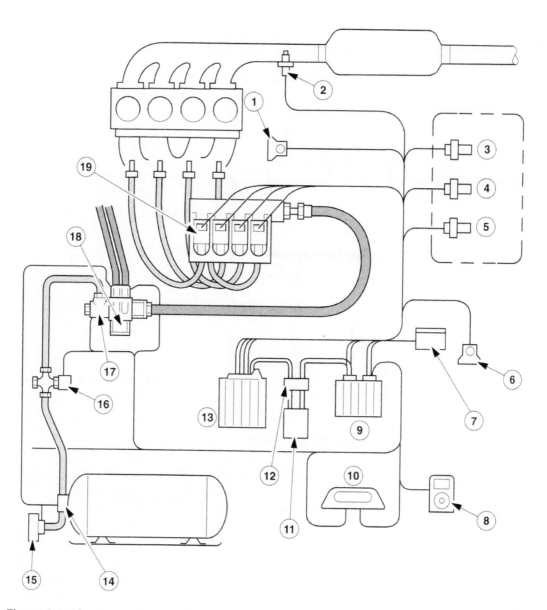

Figure 6.11 System and electronic control components: 1. Throttle position sensor, 2. Heated oxygen sensor, 3. Engine coolant temperature sensor, 4. Crankshaft position sensor, 5. Camshaft position sensor, 6. Manifold absolute pressure sensor, 7. Auxiliary fuse and relay box, 8. CNG/petrol switch, 9. Alternative fuel control module, 10. Digital fuel gauge, 11. Standard corporate protocol (SCP) translator, 12. Data link connector, 13. Power control module, 14. Main solenoid valve (fuel tank), 15. Microswitch, 16. CNG high-pressure sensor, 17. Coolant temperature sensor (in fuel pressure regulator), 18. Fuel pressure regulator solenoid valve, 19. Fuel injectors. (Source: Ford Motor Company)

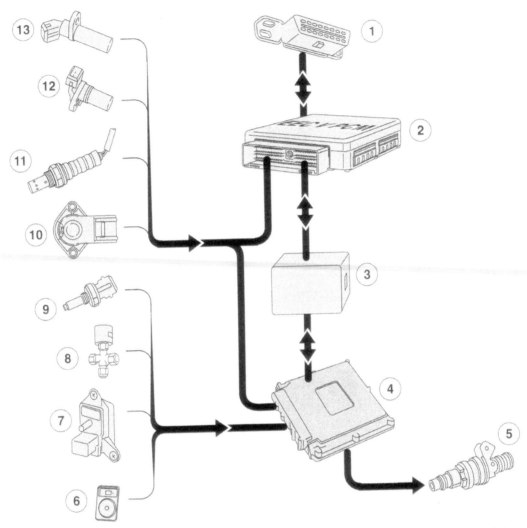

Figure 6.12 System inputs and outputs: 1. Data link connector, 2. Powertrain control module, 3. SCP translator, 4. Alternative fuel supply module, 5. Solenoid valves, 6. CNG/petrol switch, 7. Manifold absolute pressure sensor, 8. CNG high-pressure sensor, 9. Coolant temperature sensor, 10. Throttle position sensor, 11. Heated oxygen sensor, 12. Camshaft position sensor, 13. Crankshaft position (CKP) sensor. (Source: Ford Motor Company)

▶ Manifold absolute pressure sensor
▶ High-pressure fuel sensor.

The signal from the pressure sensor is used to determine fuel level in the tank. An inertia switch is incorporated into the circuit that will shut off the solenoids in the fuel tank and pressure regulator in the event of a serious impact. Diagnostics trouble codes can be accessed using the diagnostic link socket just like other vehicle systems.

The mixture is run at a lambda value of 1 (stoichiometric ratio). Because different grades of CNG exist, the ECU can make adjustments using trim values based on the lambda sensor reading. In this way it self-learns and keeps the updated information in the keep alive memory (KAM).

Definition

Stoichiometric ratio: The air to natural gas ratio by volume for complete combustion is 9.5:1 to 10:1. Note that liquid fuels are normally stated as mass ratio and gaseous fuels as a volume ratio.

6.5 Ford Taurus bi-fuel

6.5.1 Overview

In the USA the Ford Taurus was available, from several years ago, as a bi-fuel option so it would run on 100% petrol/gasoline or up to 85% methanol (M85) or ethanol (E85). The powertrain control module (PCM) can adapt to these different options. However, vehicles must be specially equipped to run on pure alcohol.

Vehicles running on alcohol can have problems starting from cold. For this reason, they can be equipped with cold starting aids such as a pre-fuel heater or heated injectors. Another option is a special fuel additive. Alcohol vehicles,

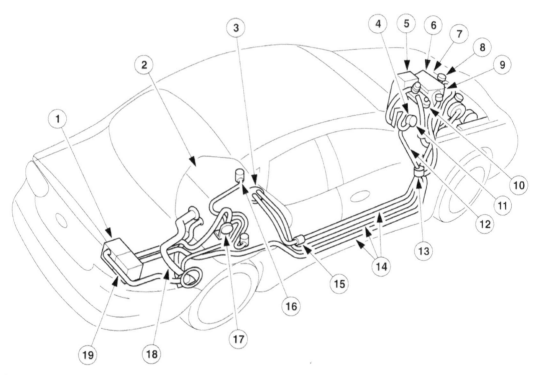

Figure 6.13 Taurus components: 1. Carbon filter, 2. Fuel tank, 3. Fuel supply module, 4. Cylinder block heater, 5. Throttle housing, 6. Intake manifold, 7. Injection valves, 8. Pressure regulator, 9. Spark plugs, 10. Exhaust gas recirculation sensor, 11. Vaporisation control valve, 12. Powertrain control module, 13. Fuel sensor and mixer, 14. Fuel and vapour pipes, 15. Fuel filter, 16. Fuel vaporisation valve, 17. Vaporisation control valve, 18. Fuel filler pipe, 19. Vaporisation filter system. (Source: Ford Motor Company)

compared with those running on petrol/ gasoline, are very similar other than a few key requirements:

> **Key fact**
>
> Vehicles running on alcohol can have problems starting from cold.

- The fuel tanks are larger because ethanol and methanol are less energy dense than petrol
- Fuel tanks are made of stainless steel as are the metal pipes to prevent corrosion
- Plastic pipes are made from a special polyamide to prevent them from swelling
- An additional fuel sensor is used to measure the mix ratio
- A fuel mixer prevents separation
- A pre-fuel heater
- Special spark plugs
- Special fuel filters
- Special engine oil prevents particles being washed from the oil that could contaminate injection valves.

6.6 Honda Clarity fuel cell

6.6.1 Overview

Honda aimed to eliminate the design compromises that are often found in alternatively fuelled vehicles. Their goal was to make a fuel cell car suited to everyday driving and long-distance travel. The Clarity currently has the longest maximum driving range rating of any zero-emission vehicle, offering approximately 650 km (403 miles) under NEDC conditions.

> **Definition**
>
> NEDC: New European driving cycle (soon to be superseded).

The fuel cell stack, which was in the centre tunnel of the previous generation model, was made significantly smaller. Even though the state-of-the-art fuel cell stack is smaller than in its predecessor, the FCX Clarity, it has a higher power output (103 kW compared with 100 kW). Innovative downsizing of the individual cells within the stack has enabled the complete unit to be housed on top of the motor, gearbox and power control unit (PCU) assembly.

More space was freed up under the bonnet by reorienting the main drive motor, gearbox and a new PCU. This now combines the power drive unit (PDU) and the battery voltage control unit (VCU), used on the earlier FCX Clarity, into a smaller housing. The overall drive motor assembly was rotated through 90 degrees, taking the assembly from a vertical to a horizontal orientation.

> **Key fact**
>
> By combining functions and reorienting the motor unit, Honda was able shrink the height of the fuel stack, control unit and drive assembly by 34%.

6.6.2 Fuel cell stack

The fuel cell stack uses 358 individual cells, which are each 20% thinner than those in its predecessor. The smaller cells are enabled by flatter and narrower flow channels for the hydrogen, oxygen and coolant, as well as a slimmer membrane electrode assembly.

Narrower gas flow paths in the separators mean that the amount of surplus condensate water in the oxygen channel is lower, which improves the air flow to the membrane. The enhanced air flow, as well as the use of more efficient membrane electrode plates, has increased the electric power output per cell by 50% compared with Honda's previous stack.

Figure 6.14 Honda Clarity. (Source: Honda Media)

Figure 6.15 Honda Clarity components: 1. Power control unit (PCU), 2. Fuel cell stack, 3. Lithium ion battery, 4. Hydrogen fuel tanks. (Source: Honda Media)

Figure 6.16 Clarity fuel cell stack, motor, compressor and control components. (Source: Honda Media)

Complementing this information for the driver, an on-demand telematics system shows real-time status information about the nearest hydrogen stations on the navigation display.

The fuel cell voltage control unit (FCVCU) delivers 500 V to the main drive motor, resulting in a 30% improvement in motor output, even while reducing the number of cells in the fuel cell stack. To create a thinner FCVCU, newly developed silicon-carbide power semiconductors are used. These can operate at a switching frequency four times higher than conventional silicon.

Definition

Silicon-carbide power semiconductors: Made from silicon and carbon with the chemical formula SiC, these electronic devices can operate at high temperature and high voltage.

The increase in power density of the cells meant that fewer of them were needed to reach the desired power output for the stack. Further packaging benefits were achieved by changing the orientation of the stack. Previously, humidity in the gas flow paths within the stack meant that the fuel cells needed to be oriented vertically to allow gravity to drain away the surplus water. Improved humidity management allows the cells, and therefore the entire stack, to be oriented horizontally.

6.6.3 Air compressor

Situated at the base of the powertrain next to the motor assembly, a two-stage supercharging electric turbo air compressor is used for the first time in a production car. While the narrower gas flow paths in each fuel cell boost the power density of the stack, they require air to be supplied at a much higher pressure. Honda designed a new electric turbo compressor that boosts the air supply pressure by 70% to achieve this.

Key fact

An increase in power density of the cells meant that fewer of them were needed to reach the desired power output for the stack.

Key fact

Narrower gas flow paths in each fuel cell mean air must be supplied at a much higher pressure.

Real-time fuel cell power use is shown by a glowing blue ball graphic, which changes size according to the energy being discharged.

The compressor is significantly quieter than the previous-generation air pump, which therefore enabled the amount of sound-suppression materials in the new car to be reduced, saving more weight. The compressor is also 40% smaller than the earlier air pump, which further benefitted Honda's goal of reducing the overall powertrain size.

6.6.4 Lithium-ion battery

A high-performance lithium-ion battery pack is accommodated below the driver and front passenger seats. It is lighter and smaller than an equivalent nickel metal hydride (NiMH) unit. Its output is 50% higher than the previous model.

This battery pack stores electricity generated by the fuel cell stack, as well as that which is generated while decelerating. It provides supplementary power to the drive motor in dynamic situations where the fuel cell stack alone would not be sufficient, such as standing starts and fast acceleration.

6.6.5 Hydrogen supply system

In the FCX Clarity, the gas pressure and flow rate from the hydrogen storage to the fuel cell stack was adjusted using a regulator relief valve and ejector. For the new Clarity Fuel Cell, Honda developed an integrated supply unit that uses two gas injectors. The new system enables a more accurate control of both the pressure and flow rate from the tanks, and occupies approximately 40% less space.

Figure 6.17 Filling with hydrogen. (Source: Honda Media)

While the overall hydrogen tank capacity in litres has been reduced compared with the previous-generation car, the contents are kept at twice the pressure (700 bar, up from 350), which enables the car to store 39% more hydrogen, up to a maximum of approximately 5 kg.

Two tanks are used, a main tank of 117 litres under the luggage compartment, and a 24-litre tank underneath the rear passenger seats. A hydrogen leak prevention and dispersion technology provides outstanding levels of safety. Sensors throughout the car constantly monitor the powertrain for potential inconsistencies in the hydrogen gas supply. If a leak is detected, the system can be shut off to prevent any more gas escaping, while precisely routed ducts allow for safe venting and dispersion of any leaked hydrogen.

Key fact
The Honda safety system allows hydrogen to be safely released from the vehicle in the event of a fire.

6.6.6 High-output drive motor

Despite having a higher power output and a faster maximum rotational speed, the main drive motor is quieter and more refined than the previous Clarity's motor. Additional structural ribs on the casing of the motor improve the rigidity of the unit and reduce vibrations. Torque fluctuations from the magnetic locking effect in the motor mechanism have been reduced, by using twice the number of skewed rotor bars. As a result, the motor sound transmitted to the interior when the car is accelerating is significantly lower.

6.6.7 Technical specifications[2]

▶ Powertrain: Polymer electrolyte fuel cell (PEFC)

- ▶ Drive motor: AC synchronous motor
- ▶ Motor output: 130 kW @ 4,501–9,028rpm (max. 13,000rpm)
- ▶ Motor torque: 300Nm @ 0–3,500rpm
- ▶ Number of cells in stack: 358
- ▶ Fuel cell output: 103 kW
- ▶ Drive battery: Lithium-ion
- ▶ Maximum speed: 165 km/h
- ▶ 0–100 km/h: 9.0s (Honda internal tests)
- ▶ Maximum range approx.: 650 km (403 miles) under NEDC conditions (Honda's internal tests).

6.7 Ford Escape hybrid E85

The Ford Escape Hybrid E85 has been around for a while now, but is an interesting combination of two fuel saving technologies:

1 Hybrid electric power
2 Flexible fuel capability.

However, this combination came with a unique set of challenges. Ethanol is more corrosive than traditional fuel. For this reason, a production level fuel/electric hybrid Escape, was retrofitted with fuel and engine system components of corrosion-resistant materials and adhesives.

Ethanol also does not possess the same energy density or burn rate as petrol/gas, which requires the system to deliver more fuel to the injectors to keep performance levels comparable. To handle this increased fuel flow, the Escape Hybrid E85 has a larger fuel pump and larger injectors.

The engine can run on conventional fuel or E85, but to achieve this the engine control module (ECM) must learn what is in the tank and make appropriate adjustments. To do this, the ECM monitors the exhaust gas sensor, the air–fuel ratio and the quantity of fuel in

Figure 6.18 Escape Hybrid E85. (Source: Ford Media)

Figure 6.19 Escape Hybrid E85 information panel. (Source: Ford Media)

the tank. When it senses a change in the engine's air–fuel ratio to the lean side (more air than fuel), it deduces that the vehicle was filled with E85 and adjusts the fuel system accordingly.

> **Key fact**
>
> The engine control module monitors the exhaust gas sensor, the air-fuel ratio and the quantity of fuel in the tank to determine what fuel is being used.

The vehicle learns while the engine is running but, because it is a hybrid, the engine may be shut off for long periods of time while the vehicle runs on electric power. For this reason, Ford had to make some custom software and calibration changes within the ECM to make sure it could remember or relearn the correct percentage of ethanol after a shutdown.

6.8 Volvo heavy trucks

Volvo is diversifying into alternative technologies that include LNG engines and hybrid drivelines for long-haul heavy vehicles. Short-range heavy trucks that use CNG are already in use so there is a network in place to distribute the fuel. The technology used for LNG, together with its superior energy density compared with CNG mean that long-haul is now possible.

> **Key fact**
>
> LNG has a superior energy density compared with CNG.

Spark ignition was considered as an option but is about 20% less efficient than compression ignition for these applications. However, gas will not work in compression ignition

Figure 6.20 Filling the Volvo FH with LNG is a simple process. (Source: Volvo Media)

engines. Volvo is therefore returning to a once discredited technology, dual-fuel, where diesel is injected and compressed to ignite the methane.

Some aftermarket conversions have been tried but diesel has formed up to 40% of the total fuel. In addition, the emission of unburnt methane through the engine was problematic. This was caused by the valve overlap. To overcome this problem, Volvo inject the gas only after the exhaust valve has closed. The diesel injected acts as a kind of spark plug. The result is the diesel only now forms about 10% of the total fuel. Natural gas is very clean burning but better still it acts as a kind of gateway fuel for the carbon-neutral biogas. The new technology is expected to be launched by 2018.

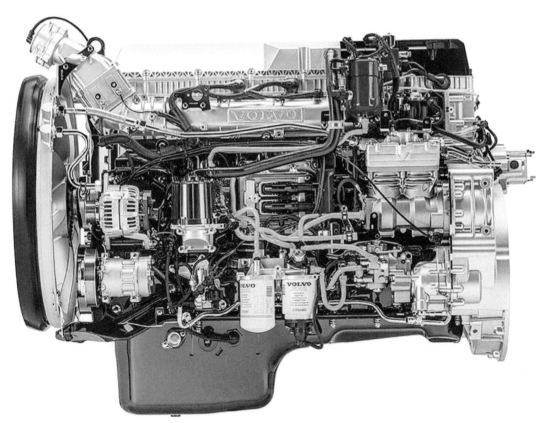

Figure 6.21 Heavy truck LNG engine. (Source: Volvo Media)

The Euro VI Volvo engine specifications are as follows:

- 13 litre, six inline cylinders
- Common rail injection
- Versions: 414bhp with 2,100Nm maximum torque, and 453bhp with 2,300Nm maximum torque
- LNG pressure 4–10bar
- LNG temperature −140 to −125°C; it is warmed before injection
- LNG tank sizes: 115kg (275L), 155kg (375L) and 205kg³ (495L).

Volvo have not stopped development of heavy-duty diesel engines but there is an interesting comparison to be made. It now takes about 100 heavy-duty diesel trucks to produce the same pollution as one similar one from 1980! At that time only one engine had to be proved to meet Euro 1 legislation, and this was done under laboratory conditions. Economy has also improved by about 1% per year since then.

> **Key fact**
> Euro 6 is reducing NOx by as much as 98%.

Volvo are continuing to work on electric propulsion, and hybrid buses are in common use. An interesting development in hybrid technology is their 'I-See' GPS-enabled predictive gearchange system. This enables a truck to adopt the optimum regeneration strategy meaning up to 30% of the driving can be done on full electric.

6.9 Volkswagen Golf dual fuel

6.9.1 Overview

The operation of this dual fuel vehicle is like the Volkswagen LPG system discussed earlier, except that this one used natural gas (CNG). The parts required for gas operation are shown in Figure 6.22.

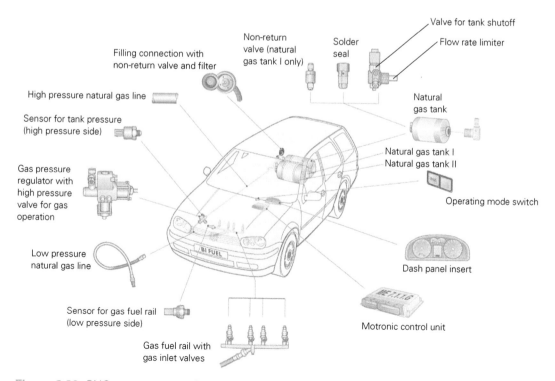

Figure 6.22 CNG components. (Source: Volkswagen)

The name and purpose of the components
shown in Figure 6.22 are outlined in Table 6.1

6.9.2 Components

Table 6.1 CNG components and their purpose

Component	Purpose and operation
Gas pressure regulator with high-pressure valve for gas operation	Reduces natural gas pressure in two stages from a maximum 200 bar to approx. 9 bar. It therefore separates the high-pressure side from the low-pressure side.
Sensor for tank pressure (high-pressure side)	Calculates the gas pressure on the high-pressure side directly before the first pressure reduction stage. Also operates the fuel gauge and its signal is needed to allow change over to gas operation.
High-pressure natural gas line	Made of stainless steel and designed for pressures above 200 bar, individual elements are joined together with a double clamping ring.
Filling connection with non-return valve and filter	Covered by a retractable cover it also includes a particulate filter and a non-return valve.
Non-return valve	Prevents accidental back flow of natural gas in the lines and to the outside via the filling connection. Two non-return valves are used, one directly on the filling connection and the other on tank shut-off valve 1 (natural gas tank 1).
Solder seal	Prevents the natural gas tank from bursting if there is an excessive increase in pressure caused by fire. The melting point of the solder is about 110°C.
Valve for tank shut off	Regulates the supply of natural gas within the vehicle.
Flow rate limiter	Prevents accidental and rapid outlet of natural gas from the tank in the event of damage to the gas lines. It is adjusted by the spring so that it closes at a pressure difference of 2 bar.
Thermal safety device	Consists of a housing into which a tube, filled with a special fluid, is fitted. The fluid expands at a temperature of 110°C and ejects itself.
Natural gas tanks	Operating pressure at 15°C, 200 bar, test pressure, 330 bar, temperature range, −40°C to +65°C.
Operating mode switch	Change between natural gas and petrol operation.
Dash panel insert	Includes the gas mode control lamp and gas fuel gauge.
Motronic control unit	Stores separate maps for both petrol operation and natural gas operation. If natural gas mode is not possible, the unit switches to petrol mode automatically. For complete combustion, a petrol engine requires a mixture of 14.7 kg of air to 1 kg of petrol, a λ value or 1. The air to natural gas ratio is 5.4:1.
Gas fuel rail with gas inlet valves	Gas is distributed to the inlet valves via fuel rail or gas rail. It enters the inlet port sequentially for each cylinder via an electronically controlled gas inlet valve at a pressure of about 9 bar (Figure 6.23).
Gas inlet valves (injectors)	The gas inlet valve is like an electromagnetic petrol injector. The valve is formed by the body and the needle that passes through it with solenoid armature. When the coil is energised, a magnetic field is generated, and the solenoid armature pulls the valve needle off its seat (Figure 6.24).
Sensor for gas fuel rail (low pressure side)	Determines the current gas pressure in the gas fuel rail. The Motronic control unit uses the signal to determine the period of gas inlet.
Low pressure natural gas line	Starts directly after the second pressure reduction stage in the gas pressure regulator. A flexible low-pressure gas line is installed from the gas pressure regulator to the gas fuel rail to allow for engine movement.

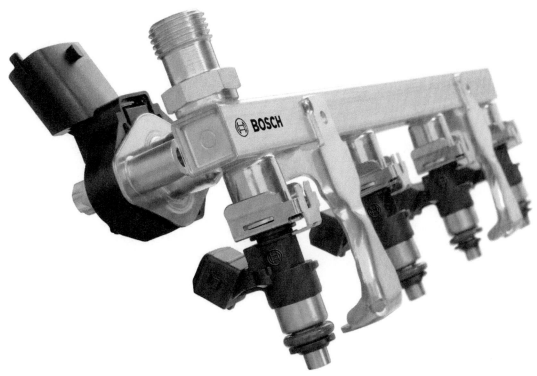

Figure 6.23 CNG fuel injectors on the rail. (Source: Bosch Media)

Figure 6.24 Natural gas injector. (Source: Bosch Media)

6.10 New developments

6.10.1 Skyactiv-X spark controlled compression ignition

A recent interesting development in combustion is Mazda's innovative Skyactiv-X spark controlled compression ignition (SpCCI) engine. It is possible it will be in production in 2018 and could extend the life of petrol/gasoline engines, which are under threat from emissions legislation and electric vehicles. SpCCI is designed for engines that will also eventually run on micro-algae biofuels, so zero net emissions are possible.

The aim of the research and development was to create an engine with the economy and torque of a CI engine, but with the higher revving capacity of a twin-cam SI engine.

93

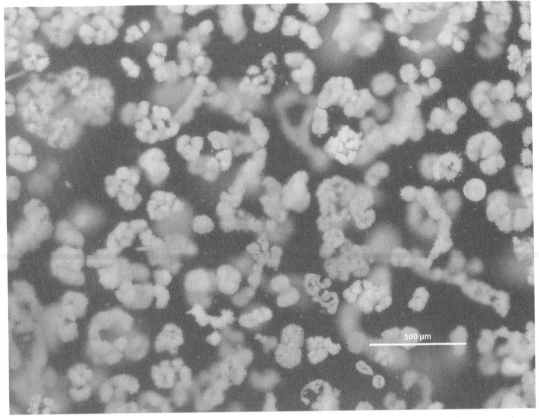

500 µm

Figure 6.25 Algae is a feedstock for future fuels. (Source: NOAA Great Lakes Environmental Research Laboratory, via Wiki Commons)

The new Skyactiv-X engine operates at a 15:1 mechanical compression ratio. However, the key aspect is that the spark creates an expanding fireball that acts like an 'air spring' to create additional compression. Because the spark plug creates this fireball, it effectively controls the switch between spark ignition and compression ignition.

Key fact

In the Skyactiv-X engine, the spark creates an expanding fireball that acts like an 'air spring' to create additional compression.

A high compression ratio is key to a lean fuel–air mixture because the leaner the air–fuel ratio, the higher the specific heat ratio. Stoichiometric levels in the region of 30:1 or more are possible.

To achieve these benefits, precise control of the combustion process is needed. The engine therefore uses sensors in each cylinder for real-time temperature and pressure monitoring. Other engine parameters such as speed and load are used, as in any engine control system. The engine management system controls twin electrically variable camshafts, a split-injection strategy that operates at 500 bar, as well as a Roots-type air blower.

Figure 6.26 Mazda engine. (Source: Mazda)

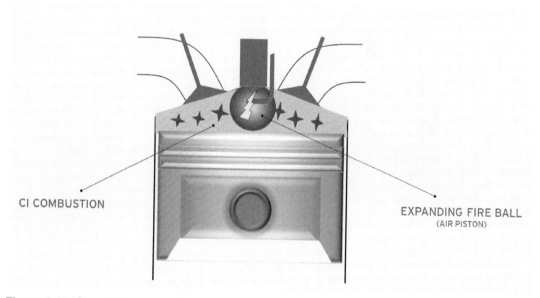

CI COMBUSTION

EXPANDING FIRE BALL
(AIR PISTON)

Figure 6.27 SI and CI are possible because of the fire ball

The air–fuel mixture is created in two phases, split between the intake and compression strokes. A strong swirl is created in the combustion chamber to ensure uneven distribution of fuel. This causes a lean mixture around the edges for CI, but a relatively rich mixture around the spark plug. This is what creates the fire ball (see Figure 6.27). Spark ignition is used to start the engine, and is used under heavy-load conditions, but the switchover to CI is not at any pre-set point. Instead, when the right intake-charge boundary conditions are achieved, the expanding fireball in the combustion chamber is created. With SI providing additional compression to the mechanical compression ratio of an additional 15 or 16:1, the CI process is started.

6.10.2 Sugar cane and jet fuel

The aviation industry produces about 2% of transport-related carbon dioxide emissions.

Electricity generation and home heating account for more than 40%, so the aviation figure is relatively small. However, it is one of the world's fastest-growing greenhouse gas sources because demand is expected to double in the next 20 years.[4] A team of researchers therefore recently started considering other options.

> **Key fact**
> The aviation industry produces about 2% of transport-related carbon dioxide emissions.

Bio-jet fuels derived from oil rich feedstocks, such as algae, have been successfully tested in proof of concept flights. A global standards organisation has approved a 50:50 blend of petroleum-based jet fuel and hydro-processed renewable jet fuel for normal use.[5] However, current production volumes of bio-jet fuel

Figure 6.28 Harvesting sugarcane in Brazil. Jonathan Wilkins, CC BY-SA

are very small, because making these products on a larger scale requires technology improvements and access to large quantities of low-cost feedstocks.

Sugarcane is a well-known biofuel source. For example, Brazil has been fermenting sugarcane juice to make alcohol-based fuel for decades. Ethanol from sugarcane yields 25% more energy than the amount used during the production process, and therefore reduces greenhouse gas emissions significantly.

The research team wondered if it would be possible to increase the sugar cane's natural oil production and use the oil to produce biodiesel. Biodiesel yields over 90% more energy than is required to make it and reduces emissions by 41% compared with fossil fuels.

Key fact
Biodiesel yields over 90% more energy than is required to make it.

Normal sugarcane plants contain just 0.05% oil, which is far too little to convert to biodiesel. In addition, many plant scientists

theorised that increasing the amount of oil to just 1% would be toxic to the plant. None-the-less, the team used computer models to predict that they could increase oil production to 20%. Since then, through genetic engineering, they have increased production of oil and fatty acids to achieve 12% oil in the leaves of sugarcane (Kumar, Long, & Singh, 2017).

Notes

1 I am grateful to Volkswagen, Ford, Toyota, Honda, Volvo and many other companies for the information used in this chapter.
2 Technical information is preliminary and subject to change and fuel consumption figures are subject to final homologation.
3 Has a claimed operating range of 1,000 km.
4 Source: http://theconversation.com/jet-fuel-from-sugarcane-its-not-a-flight-of-fancy-84493
5 Source: https://www.astm.org/cms/drupal-7.51/newsroom/astm-aviation-fuel-standard-now-specifies-bioderived-components

CHAPTER 7

Automotive Technology Academy

7.1 Introduction

The online Automotive Technology Academy has been created by the author (Tom Denton), who has over 40 years of relevant automotive experience, and over 20 published textbooks that are used by students and technicians worldwide.

The aims of the online academy are to:

▶ Improve automotive technology **skills** and **knowledge**
▶ Provide **free** access to study resources to support the textbooks
▶ Create a worldwide **community** of automotive learners
▶ Freely **share** automotive related information and ideas
▶ **Reach** out to learners who are not able to attend school or college
▶ Improve automotive training **standards** and **quality**
▶ Provide online access to **certification** for a range of automotive subjects.

To access the academy visit: www. automotive-technology.org and create an account for yourself. To access the free courses, which work in conjunction with the textbooks, you will need to enter an enrolment key. This will be described something like this: 'The third word on the last line of page ## of the associated textbook.' You will therefore need to own the book! All you need to do is enter the word in a box and you will have full, unrestricted access to the course and its associated resources.

Website

www.automotive-technology.org

7.2 Resources

The following is a list of some of the resources that will be available to you:

▶ Images
▶ Videos
▶ Activities
▶ 3d models
▶ Hyperlinks
▶ Assignments

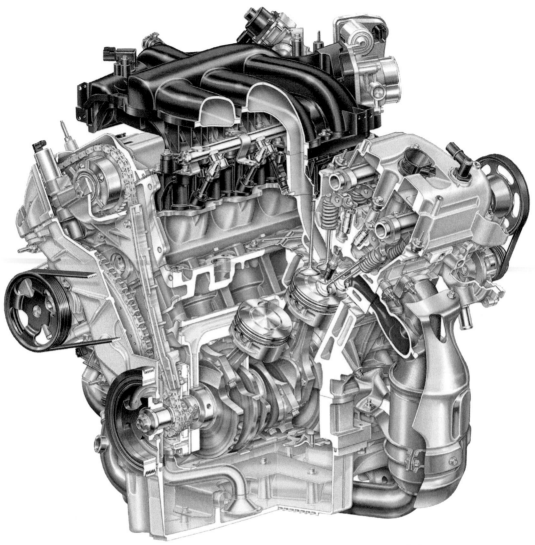

Figure 7.1 Four-cylinder double overhead cam engine. (Source: Ford Media)

▶ Quizzes
▶ Forums
▶ Chat features
▶ Social media
▶ Interactive features, games, and much more.

A progress bar is used in the courses, so you can see at a glance how you are getting on, if you are working to cover all the content. Alternatively, you can just dip in and out to find what you need.

More formal assessments will also be available for those who are not able to attend traditional training centres. It will be possible to obtain certification relating to theory and practical work. A charge will apply to this aspect, but all other resources are free.

Updates and interesting new articles will also be available, so what are you waiting for? Come and visit and join in!

Figure 7.2 Biodiesel engine

References

BP. (2017). Statistical Review of World Energy. BP.

Denton, T. (2016). *Advanced Automotive Fault Diagnosis* (4th ed.). London and New York: Routledge.

Denton, T. (2016). *Electric and Hybrid Vehicles*. London and New York: Routledge.

Kumar, D., Long, S., & Singh, V. (2017). Biorefinery for combined production of jet fuel and ethanol from lipid-producing sugarcane: a techno-economic evaluation. *GCB Bioenergy*. doi:10.1111/gcbb.12478.

Index

Pages numbers in *italics* denote figures, those in **bold** denote tables.

Printed and bound by CPI Group (UK) Ltd, Croydon, CR0 4YY

21/10/2024

01777046-0003